Mechanisms of Secondary
Brain Damage in Cerebral
Ischemia and Trauma

Edited by
A. Baethmann, O. Kempski,
N. Plesnila, and F. Staub

Acta Neurochirurgica
Supplement 66

SpringerWienNewYork

Professor Dr. Alexander Baethmann
Institut für Chirurgische Forschung, Ludwig-Maximilians-Universität, Klinikum Grosshadern, München,
Federal Republic of Germany

Professor Dr. Oliver S. Kempski
Institut für Neurochirurgische Pathophysiologie, Johannes-Gutenberg-Universität, Mainz,
Federal Republic of Germany

Dr. Nikolaus Plesnila
Institut für Chirurgische Forschung, Ludwig-Maximilians-Universität, Klinikum Grosshadern, München,
Federal Republic of Germany

Dr. Frank Staub
Klinik für Neurochirurgie der Universität zu Köln, Köln, Federal Republic of Germany

© 1996 Springer-Verlag/Wien
Softcover reprint of the hardcover 1st edition 1996

Product Liability: The publisher can give no guarantee for information about drug dosage and application thereof contained in this book. In every individual case the respective user must check its accuracy by consulting other pharmaceutical literature. The use of registered names, trademarks, etc. in this publication does not imply, even in the absence of specific statement, that such names are exempt from the relevant protective laws and regulations and therefore free for general use.

Typesetting: Thomson Press, New Delhi, India

Graphic design: Ecke Bonk

Printed on acid-free and chlorine free bleached paper

With 52 Figures

Library of Congress Cataloging-in-Publication Data

Mechanisms of secondary brain damage in cerebral ischemia and trauma /
    edited by A. Baethmann ... [et al.].
        p.   cm. -- (Acta neurochirurgica. Supplementum ; 66)
    Includes bibliographical references and index.
    ISBN-13: 978-3-7091-9467-6   e-ISBN-13: 978-3-7091-9465-2
    DOI: 10.1007/978-3-7091-9465-2
    1. Cerebral ischemia. 2. Brain damage.   I. Baethmann, A.
II. Series.
RC388.5.M436   1996
616.8'047--dc20
                                                        96-3928
                                                        CIP

ISSN 0065-1419
ISBN-13: 978-3-7091-9467-6

Preface

The publication of the Vth International Symposium 1995 on "Mechanisms of Secondary Brain Damage" in Mauls/Italy is a collection of focused reviews reaching from novel molecular- and cell biological findings to aspects of clinical management in head injury and cerebral ischemia. A specific purpose of these series of meetings introduced in 1984 is for an exchange on problems of mutual interest by international high ranking experts from the basic sciences and related clinical disciplines, such as intensive care medicine, neurology, or neurosurgery. The present volume covers three major areas:

(a) Molecular and cell biological mechanisms including inflammation
(b) Novel findings on mechanisms and treatment in cerebral ischemia
(c) Secondary processes in head injury, regeneration and treatment

Molecular- and cell biology is currently attracting attention towards activation of genomic processes associated with the demise of cells referred to as "programmed cell death" and "apoptosis" which, actually, might be distinguished from each other. Thus, the phenomenon of delayed neuronal death in selectively vulnerable brain areas following brief interruption of blood flow is scrutinized as to the contribution of the activation of suicide genes. The physiological role of such a response, among others, is removal of surplus neurons during ontogenesis of the brain. Yet, evidence is accumulating that similar mechanisms play a role in cerebral ischemia, probably also trauma, where nerve- and other cells demonstrate features of apoptosis. Observations on protection of neurons by administration of protein synthesis inhibitors in cerebral ischemia provide more direct support. Expression of respective immediate early genes has been explored, indicating, that for example c-fos, c-jun, or others, such as mkp-1 are involved. A clear answer of whether or not the loss of neurons from ischemia or trauma is attributable to such controlled biological processes or just resulting from gross injury leading to cell necrosis is clinically important, because therapeutical opportunities might thereby emerge.

Other chapters are dealing with the role of acidosis and intracellular accumulation of Ca^{++}-ions in the above processes. Interestingly, an intracellular Ca^{++}-overload might not always represent the final pathway to cell destruction, as recent findings in cerebral ischemia indicate a discrepancy between the distribution of brain tissue areas with increased Ca^{++}-accumulation and those with structural damage. As to the role of inflammation, it is increasingly obvious that leukocyte/endothelial interactions enhance tissue damage from ischemia. This may also hold for the brain, in ischemic infarction, where monoclonal antibodies are available against surface molecules of endothelium and leukocytes, which mediate respective interactions. The utilization of such antibodies not only is helpful to enhance understanding of underlying molecular mechanisms but also for treatment with promising results. The publication further reports on advances to quantitatively analyze the topographic distribution of metabolic and hemodynamic changes in focal ischemia with an unmatched resolution. In addition, the significance of glutamate overflow in damaged brain tissue, particularly its source of release and the role of receptor subtypes, such as the AMPA-moiety remain in the focus of scientific and clinical efforts with growing therapeutical success.

Ultrastructural, pathophysiological and biochemical phenomena are also explored in head injury with impressive subtlety. Findings indicate that diffuse axonal injury, so far considered as manifestation of primary damage, might constitute a secondary process raising the possibility for therapeutical interventions. This holds also for the activation of microglial cells which are intensively studied in brain trauma due to their capacity of releasing powerful cytokines and mediator compounds. Further, in view of potential associations between Alzheimer's disease and head injury new data are available on the ß-amyloid precursor protein which seem to tighten this link.

An illustrative secondary process in brain trauma is the motoneuron response in the brain following axotomy, eventually leading to neuronal degeneration and involving a host of molecular and cell biological reactions. Respective studies provide evidence that the secondary degeneration of these nerve cells can be prevented by supplementation of neurotrophic factors.

Recent developments at the treatment front expand the current repertoire for cerebral ischemia and trauma. The spectrum reaches from novel antioxidative compounds penetrating the blood-brain barrier, to attempts of grafting fetal brain tissue, or enrichment of the environment to enhance the sensory input. Whereas the efficacy of fetal tissue grafting in cerebral infarction so far is disappointing, the transplant seems to survive and becomes innervated. Efforts with a particularly stimulatory environment are more successful to preserve or improve functional competence, even though infarct size is not affected. Hence, progress in treatment is accomplished not only by pharmacological compounds, but also by improvement in the general management. A case in point is the timely induction of thrombolysis in acute cerebral infarction as long as the window of opportunity is open, or the prompt reestablishment of cardiovascular competence in head injury patients with simultanous hemorrhagic shock by "small volume resuscitation" with hyperosmotic/hyperoncotic solutions.

Finally, in view of the complex requirements of logistics and management including more specific treatment modalities, activities are scaled up at an international level to advance procedures of testing new protocols for the preclinical and clinical care of patients and to obtain more reliable data which is useful for a comparison between centers. Noticeable efforts in this context are made by the American and European Brain Injury Consortium (ABIC, EBIC). Obviously, improvements in outcome from cerebral ischemia and severe head injury can preferably be expected, if the basic sciences and clinical experience are better coordinated – a fascinating as well as challenging endeavour.

We would like to thank the sponsors, Adrianus van de Roemer, and Stan Snowball, Upjohn, who made possible the publication of the symposium and – the late Hans Stafler – and his family for providing an unrivaled hospitable ambience in the Albergo Stafler in Mauls as a particular fertile ground for frank and spirited discussions. It is also a great pleasure to acknowledge support obtained by Raimund Petri-Wieder, Springer, Vienna, and the secretarial assistance of Helga Kleylein and Edith Martin.

Munich, Mainz, and Cologne, March 1996

A. Baethmann
O. Kempski
N. Plesnila
F. Staub

Contents

Molecular and Cell Biological Mechanisms – Inflammatory Processes

Wiessner, C., Vogel, P., Neumann-Haefelin, T., Hossmann, K.-A.: Molecular Correlates of Delayed Neuronal Death Following Transient Forebrain Ischemia in the Rat 1

Siesjö, B. K., Katsura, K. I., Kristián, T., Li, P.-A., Siesjö, P.: Molecular Mechanisms of Acidosis-Mediated Damage 8

Linde, R., Laursen, H., Hansen, A. J.: Is Calcium Accumulation Post-Injury an Indicator of Cell Damage? 15

Chopp, M., Li, Y.: Apoptosis in Focal Cerebral Ischemia 21

Hallenbeck, J. M.: Significance of the Inflammatory Response in Brain Ischemia 27

Tomita, M., Fukuuchi, Y.: Leukocytes, Macrophages, and Secondary Brain Damage Following Cerebral Ischemia 32

Kogure, K., Yamasaki, Y., Matsuo, Y., Kato, H., Onodera, H.: Inflammation of the Brain after Ischemia 40

Novel Findings on Mechanisms and Treatment of Cerebral Ischemia

Ginsberg, M. D., Back, T., Zhao, W.: Three-Dimensional Metabolic and Hemodynamic Imaging of the Normal and Ischemic Rat Brain 44

Obrenovitch, T. P.: Origins of Glutamate Release in Ischemia 50

Staub, F., Winkler, A., Haberstok, J., Plesnila, N., Peters, J.,Chang, R. C. C., Kempski, O., Baethmann, A.: Swelling, Intracellular Acidosis, and Damage of Glial Cells 56

Johansson, B. B.: Environmental Influence on Outcome After Experimental Brain Infarction 63

Grabowski, M., Johansson, B. B., Brundin, P.: Fetal Neocortical Grafts Placed in Brain Infarcts Do Not Improve Paw-Reaching Deficits in Adult Spontaneously Hypertensive Rats 68

Diemer, N. H., Balchen, T., Bruhn, T., Christensen, T., Vanicky, I., Nielsen, M., Johansen, F. F.: Extended Studies on the Effect of Glutamate Antagonists on Ischemic CA-1 Damage 73

Fieschi, C., Cavalletti, C., Toni, D., Fiorelli, M., Sacchetti, M. L., De Michele, M., Gori, M. C., Montinaro, E., Argentino, C.: Thrombolysis in Acute Ischemic Stroke 76

Recent Observations on Mechanisms, Regeneration, and Treatment in Head Injury

Povlishock, J. T., Pettus, E. H.: Traumatically Induced Axonal Damage: Evidence for Enduring Changes in Axolemmal Permeability with Associated Cytoskeletal Change 81

Engel, S., Wehner, H. D., Meyermann, R.: Expression of Microglial Markers in the Human CNS After Closed Head Injury 87

Graham, D. I., Gentleman, S. M., Nicoll, J. A. R., Royston, M. C., McKenzie, J. E., Roberts, G. W., Griffin, W. S. T.: Altered ß-APP Metabolism After Head Injury and its Relationship to the Aetiology of Alzheimer's Disease 96

Kreutzberg, G. W.: Principles of Neuronal Regeneration 103

Hall, E. D., Andrus, P. K., Smith, S. L., Oostveen, J. A., Scherch, H. M., Lutzke, B. S., Raub, T. J., Sawada, G. A., Palmer, J. R., Banitt, L. S., Tustin, J. S., Belonga, K. L., Ayer, D. E., Bundy, G. L.: Neuroprotective Efficacy of Microvascularly-Localized Versus Brain-Penetrating Antioxidants 107

Kempski, O., Obert, C., Mainka, T., Heimann, A., Strecker, U.: "Small Volume Resuscitation" as Treatment
    of Cerebral Blood Flow Disturbances and Increased ICP in Trauma and Ischemia 114
Marmarou, A.: Conduct of Head Injury Trials in the United States: The American Brain Injury Consortium
    (ABIC) 118

Index of Keywords 122

Listed in Current Contents

Acta Neurochir (1996) [Suppl] 66: 1–7

Molecular Correlates of Delayed Neuronal Death Following Transient Forebrain Ischemia in the Rat

C. Wiessner, P. Vogel, T. Neumann-Haefelin, and **K.-A. Hossmann**

Max-Planck-Institute for Neurological Research, Department for Experimental Neurology, Cologne and Neurologische Klinik, Heinrich-Heine Universität, Dusseldorf, Federal Republic of Germany

Summary

Following transient forebrain ischemia selective and delayed neuronal degeneration occurs in the CA1 sector of the hippocampus. It is presently unclear whether this cell death is related to programmed cell death (PCD), which occurs in neurons during development of the CNS. Recently, the expression of various genes, such as c-fos, c-jun, mkp-1, cyclin D1, and hsp70 was found to be associated with PCD in model systems. We and others have described that these genes are also upregulated in the hippocampus following ischemia. Most notably, c-fos, c-jun, and hsp70 are expressed specifically in CA1 neurons at survival times shortly preceding cell degeneration in rat models of global ischemia. In addition, the gene products could be detected by immunohistochemical methods, despite a general impairment of protein synthesis. These findings are especially relevant, since a recent report suggests a functional role for Fos family proteins and c-jun in PCD in neurons of the superior cervical ganglion. These results could be indicative for the occurrence of a PCD-related program in CA1 neurons and corroborate several other lines of evidences, such as occurrence of DNA fragmentation. Clearly, further studies are necessary to elucidate the functional role of the gene inductions following ischemia *in vivo*.

Keywords: Gene expression; programmed cell death; hippocampus; cerebral ischemia.

Introduction

Ischemic brain insults can result in alterations of genomic programme in various cell types of the CNS [23,39–41]. Gene inductions also accompany delayed neuronal death in the hippocampus after ischemia and can, therefore, be viewed as molecular correlates of neuronal death. The functional relevance for most of the investigated gene inductions, however, has not yet been fully elucidated.

Recently, a close association of gene inductions and programmed cell death (PCD) has been described in several well characterised paradigms of PCD (Table 1)

[4,5,7,10,12,32]. PCD is well known to occur in neurons during development of the central nervous system [24]. Therefore, in principle this naturally occurring death programme is intrinsic to each neuron and could be activated under pathological conditions. Such PCD-triggering events could be disturbances of calcium homeostasis, free radical formation, and impairment of protein synthesis, all of which are well known to occur after cerebral ischemia [16,29,30], and have also been implicated in PCD [8,20].

The most relevant and best characterised paradigm for PCD in neurons is that of cultured neurons from the superior cervical ganglion upon nerve growth factor (NGF) withdrawal [8,10,12]. For this paradigm, the 'killer protein' hypothesis has been put forward. This idea was initially based on the observation that protein synthesis inhibitors, if given at the appropriate time, can prevent completion of PCD. Recently, in this system the expression of more than 100 genes has been investigated, and several were found to be induced in a fashion suggesting an involvement in PCD. Among them are several immediate early genes (IEGs) such as members of the Fos family, c-jun, and mkp-1 (3CH134) which encodes a dual specificity protein phosphatase [34]. In addition, induction of cyclin D1 – a regulator of cell division – and the extracellular proteases transin and collagenase was observed. For Fos family proteins and c-jun it could be demonstrated that micro-injection of antibodies could block PCD, indicating a functional role for at least one of the Fos family proteins and c-jun in PCD in SCG neurons. Recently, we and others have investigated the postischemic expression profiles of several genes, which have

Table 1. *Gene Inductions in Paradigms of Programmed Cell Death (PCD)*

PCD paradigm	Induced genes	Reference
Cultured SCG neurons upon NGF deprival	c-fos, c-jun, c-myb, fosB, junB, mkp-1, cyclin D1, transin, collagenase (not c-myc, hsp70, sgp-2)	Freeman et al., 1994 Estus et al., 1994
Neurons in the developing CNS	c-fos	Smeyne et al., 1993
Lymphoid cells deprived of growth factors	c-fos, c-jun	Colotta et al., 1992
Prostate regression	c-fos, c-myc, hsp70, sgp-2 (=Trpm 2, Apolipoprotein J)	Buttyan et al., 1988 Buttyan et al., 1989

been reported to be induced in PCD paradigms (Table 1), such as c-fos, c-jun, hsp70, mkp-1, and sgp-2. However, these expression profiles have not yet been comprehensively discussed in light of the hypothesis of the possible activation of PCD. Therefore, in this communication we have attempted to reconcile our data with this hypothesis. We discuss the data in relationship to results obtained in other model of global ischemia and to preliminary data of recently investigated markers of PCD in our model.

Material and Methods

Four-Vessel Occlusion [25,28]

Male Wistar rats (body weight between 250 and 300 g) were anaesthetised with halothane/nitrous oxide (1 %/70 % respectively). On the first day both vertebral arteries were coagulated. On the following day, both carotid arteries were occluded with aneurysm clips for 30 min under continuous EEG monitoring. Body temperature was maintained at 37 °C using a heating pad. Brain temperature was monitored, but not kept constant during ischemia, because this requires heating of the head, which introduces a complicating factor and aggravates the ischemic injury. Only animals with complete EEG flattening shortly after occluding the carotid arteries and without convulsions in the survival period were used in the study. Four animals per time point were decapitated under deep anaesthesia at 15 min, 30 min, 1 hour, 6 hours, 12 hours, 1 day, 2 days, 3 days, and 7 days after recirculation and the brains were frozen in isopentane at −70 °C. Cryostat sections (10 μm) were thawmounted on poly-lysine coated glass-slides and fixed for 15 minutes in 4 % paraformaldehyde/phosphate buffered saline. In each animal the extent of neuronal damage was evaluated by histology using cresyl violet (Nissl) staining.

Probes

Oligonucleotide probes were radioactively labelled with terminal deoxynucleotidyltransferase (Life Technologies, Eggenstein,

Germany) and ^{35}S-dATP (1200 Ci/mmol) to a specific activity of >0.5 × 10^9 dpm/μg. C-fos, c-jun, mkp-1, hsp70, and sgp-2 probes have been described previously [22,39,43], the transin probe was complementary to positions 237–266 of rat transin-2 [2].

In situ Hybridisation

In situ hybridisation with oligonucleotide probes was performed as described previously [22]. Briefly, tissue sections were acetylated with acetic anhydride in triethanolamine. Following dehydration 10 μl hybridisation buffer (2 × SSC, 50 % formamide, 10 % dextrane sulphate, 100 μg/ml poly-A, 120 μg/ml heparin, 1 mg/ml herring sperm DNA, 5 mM DTT, 1 mg/ml BSA) containing 3–20 pg/μl oligodeoxynucleotide probe was applied to one brain slice and covered with a coverslip. Following overnight hybridisation at 42 °C (40 °C for hsp70), slides were washed in 2 × SSC/50 % formamide for 1 hour at 42 °C (40 °C for hsp70). Hybridised radioactivity was visualised by film autoradiography (Amersham hyperfilm ß-max) with exposure times of 1–3 weeks and/or dipping the sections into photo emulsion (Amersham, LM-1). In order to assure specific hybridisation for all probes, tissue sections were (a) incubated with a 100fold excess of unlabelled probe, or (b) pre-treated with RNase A (20 μg/ml, 45 min), or (c) hybridised with a 45mer non-sense oligonucleotide. These experiments did not reveal any detectable signals, thereby confirming the specificity of detection.

Results and Discussion

Following transient forebrain ischemia produced by four-vessel occlusion for 30 minutes, typical delayed neuronal death of hippocampal CA1 neurons became morphologically visible between 2 and 3 days after ischemia (Table 2). It is important to point out that all animals with survival periods of 3 days or longer showed pronounced CA1 neuronal damage, whereas at 2 days after ischemia CA1 neurons were morphologically intact in some animals. Therefore, it can be assumed that neurons with intact morphology at 2 days after the insult represent neurons at a rather late stage before delayed death in the hippocampus becomes apparent.

The comparison of the gene expression profiles following transient forebrain ischemia with that in PCD models is complicated by a generalised genomic stress response [41]. This response includes a very early induction of immediate early genes, such as c-fos, c-jun, and mkp 1 (Table 2), which is already obvious at 15 or 30 minutes after ischemia. In addition, the mRNA coding for the inducible member of the heat shock protein family, hsp70, was induced with a similar temporal and spatial pattern. This response was observed in all previously ischemic brain cells [22,41,43] and showed no topical correlation to selective vulnerability of neuronal sub-populations. This is highlighted in Fig. 1 for c-fos and c-jun, showing that the relative

Table 2. *Induction of PCD-Related Genes in the Hippocampus after Ischemia*

Survival time	CA1 damage (+)	mRNAs above control level (+) in the *stratum pyramidale* of the hippocampal CA1 sector after 30 minutes 4-vessel occlusion in the rat						
		c-fos	c-jun	mkp-1	hsp70	cyclin D1	sgp-2	transin
15 min (n=4)	–	–	–	+	–	–	–	–
30 min (n=4)	–	+	+	+	+	–	–	–
1 hour (n=4)	–	+	+	+	+	–	–	–
3 hours (n=4)	–	+	+	+	+	–	–	–
6 hours (n=4)	–	+	+	+	+	–	–	–
12 hours (n=4)	–	+	+	+	+	–	–	–
1 day (n=4)	–	+	+	+	+	–	–	–
2 days (n=2)	–	+	+	–	+	–	–	–
2 days (n=2)	+	–	–	–	–	+	–	–
3 days (n=4)	+	–	–	–	–	+	–	–
7 days (n=4)	+	–	–	–	–	–	+	–

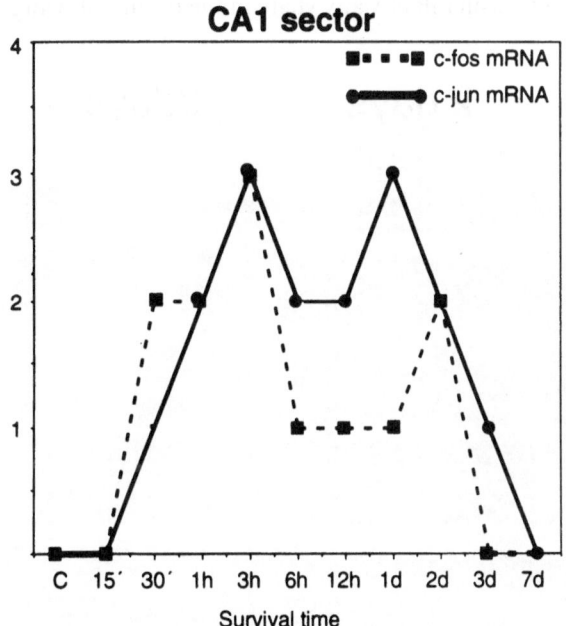

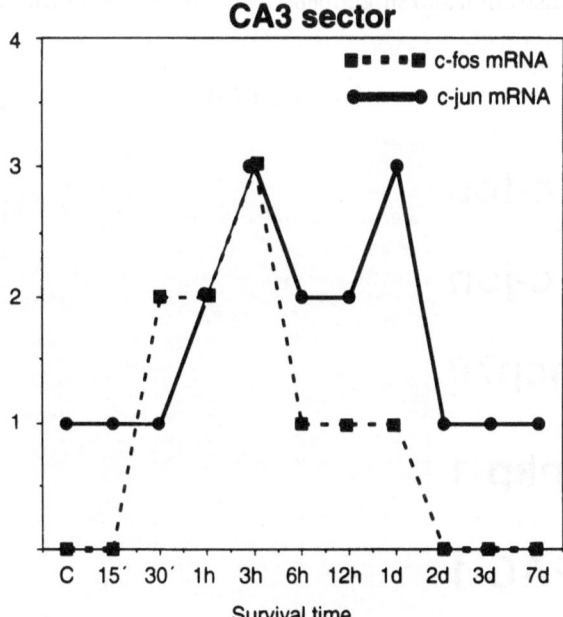

Fig. 1. Relative mRNA levels of c-fos and c-jun in the hippocampal CA1 and CA3 sector. Film autoradiograms of in situ hybridisation experiments (n=4 for each survival time) were visually inspected by a blinded investigator. The intensities of the hybridisation signals in the stratum pyramidale of the CA1 sector were qualitatively grouped from 0 (no signal) to 4 (very strong signal). Note the peak at 3 hours for both c-fos and c-jun mRNA levels in both the CA1 and CA3 sector, and the CA1 specific expression for c-fos and c-jun at 2 days after ischemia. Note also the difference of basal c-jun mRNA levels between CA1 and CA3

mRNA levels peak at 3 hours after ischemia in CA1 and CA3 neurons.

Most interestingly, at 2 days after ischemia, a CA1 specific expression was observed for c-fos and c-jun (Table 2, Fig. 2) in animals without morphological damage of the CA1 neurons. Once, neuronal damage became obvious in the Nissl staining, the mRNA was no longer detectable, which is most likely the result of rapid degradation (Table 2).

Elevated c-fos and c-jun mRNA levels in the CA1 sector at times shortly preceding cell degeneration have also been described in other experimental models of transient forebrain ischemia in the rat and following hypoxic/ischemic insults [9,17,37]. These results are quite similar to the findings in SCG neurons undergoing PCD [10]. In this model a co-ordinated expression of c-fos and c-jun was found 15 hours after onset of NGF withdrawal, i.e. shortly before the onset of cell degeneration. Another gene for which a late CA1 specific expression was found is hsp70 (Table 2, Fig. 2). This persistent CA1 specific expression for hsp70 has been reported for most global ischemia models [23]. An expression of hsp70 shortly before cell death has also been reported for epithelial cells of the rat prostate gland, which die by programmed cell death following castration and subsequent depletion of male hormones

[4]. The late up-regulation of hsp70 mRNA was not found in SCG ganglion neurons during PCD [10]. Another gene expressed in SCG neurons undergoing PCD, mkp-1 (Table 1), was also induced in the CA1 sector (Table 2, Fig. 2) following ischemia. Mkp-1 showed a similar early induction in the hippocampus when compared with c-fos and c-jun (Table 2). The mRNA was found in hippocampal neurons up to 24 hours after ischemia, showing no topical correlation to selective vulnerability. For this gene the CA1 specific expression at 2 days after ischemia was not observed (Table 2, Fig. 2). This finding is again similar to the SCG neuron paradigm of PCD, where mkp-1 mRNA levels declined earlier than c-fos and c-jun mRNAs [10].

Recently we started to investigate cyclin D1 expression, which is induced in the SCG paradigm of PCD (Table 1) with kinetics similar to those of c-fos and c-jun. Preliminary results in our model of transient forebrain ischemia show that cyclin D1 is also induced, but the onset of induction lagged behind c-fos, c-jun and hsp70 (Wiessner, in preparation). In general, cyclin D1 mRNA was found exclusively in brain regions showing morphological damage of neurons, raising the question of the cell type expressing cyclin D1. We could not detect transin mRNA anywhere in the brain following

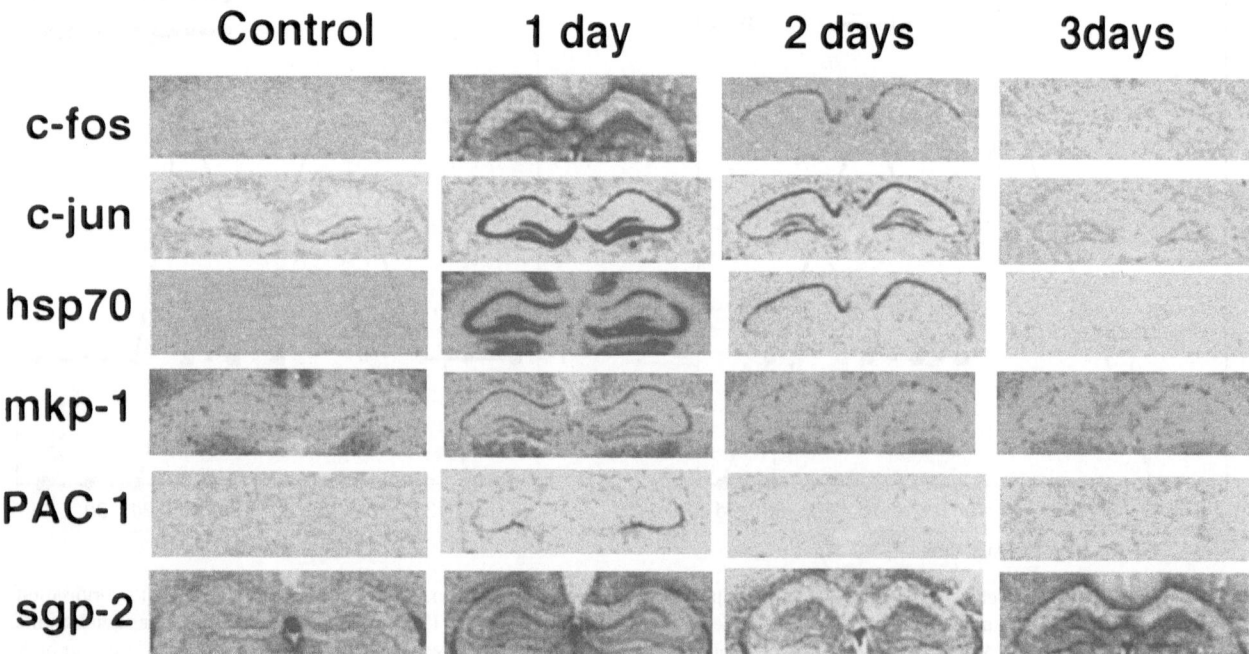

Fig. 2. Hippocampal expression of several genes in control and postischemic rat brains at various times points after 30 min of forebrain ischemia as indicated in the figure. The figure shows digitalized film autoradiograms obtained by in situ hybridisation experiments with coronal brain sections at the level of the dorsal hippocampus. Note that blackening of the films represents a relative measure indicating increases for individual mRNAs, but does not allow comparison of absolute mRNA levels between different mRNAs. No apparent morphological damage of the hippocampus was observed in the shown animal with 2 days survival period (s. also Table 2). Note the CA1-specific increased mRNA levels for c-fos, c-jun, and hsp70 at 2 days after ischemia, and the CA3 specific expression of the immediate early gene PAC-1 at 1 day after ischemia

global ischemia (Table 2). In SCG neurons undergoing PCD, transin was induced, although delayed as compared to the immediate early genes c-fos, c-jun, and mkp-1 and cyclin D1. In a recent study, we and others investigated the expression profile of sgp-2 mRNA following transient forebrain ischemia [26,39,44], which is induced in prostate gland cells during PCD, induced by depletion of male hormones [5]. Following forebrain ischemia, induction of this gene in the hippocampus was observed from 12 hours after ischemia onwards. In the stratum pyramidale of the CA1 sector, however, increased mRNA levels were found only at 7 days after the insult (Table 2). The combination of in situ hybridisation with the sgp-2 mRNA probe and immunohistochemical staining of astrocytes with an GFAP (glial fibrillary acidic protein) antibody revealed that sgp-2 mRNA was induced mainly in reactive astrocytes [39]. This finding excluded an association of neuronal cell death and expression of this gene in neurons and prompted us to conclude that sgp-2 induction is more likely important for astrocytic responses in ischemia. These observations show that the expression profiles of several PCD-associated genes (Table 1), such as cyclin D1, transin, and sgp-2 were different between the postischemic hippocampus and PCD-models.

In conclusion, the CA1 specific expression of c-fos, c-jun, and hsp70 after transient forebrain ischemia at times shortly preceding cell degeneration (Table 2) might be suggestive for the ultimate activation of a cell death programme in these neurons. In addition, the induction of mkp-1 is similar to SCG neurons undergoing PCD. Since all these genes are also expressed in resistant CA3 neurons at earlier reperfusion periods

(Fig. 2, Table 2), additional regulatory mechanisms must be postulated, which would differentiate CA1 from CA3 neurons. Such a mechanism could be the postischemic induction of the immediate early gene PAC-1, which was found in CA3 neurons and granule cells of the dentate gyrus, but not in CA1 neurons (Fig. 2) [42]. This IEG encodes a dual specificity phosphatase, which might be important for the postischemic balance of intracellular signalling cascades. An important point is, that in rat models of global ischemia the protein products for c-fos, c-jun, and hsp70 could be detected by immunohistochemical methods (shown for hsp70 in Fig. 3) [22,35]. This is especially noteworthy, since it is well documented that global protein synthesis rates are severely and permanently impaired in the CA1 sector following 30 minutes transient forebrain ischemia in the rat [38]. Therefore, despite a general impairment of protein synthesis, the cells are obviously able to translate specific mRNAs into proteins. Interestingly, this accumulation was not found in gerbil models of global cerebral ischemia [18,23]. It is unknown whether this reflects a species difference or is due to different time courses of events, i.e. although expression of PCD-related genes occurs also in the gerbil it might have been missed by the choice of time points of investigation.

Several other lines of evidence suggest a relationship of delayed neuronal death to programmed cell death. A recent study reports the specific activation of p53 and p21 (WAF1) in the CA1 sector following transient forebrain ischemia in the rat [33]. p53 is a transcription factor for p21, an inhibitor of cyclin dependent kinases. Expression of p53 is known to be involved in the initiation of programmed cell death in various cell

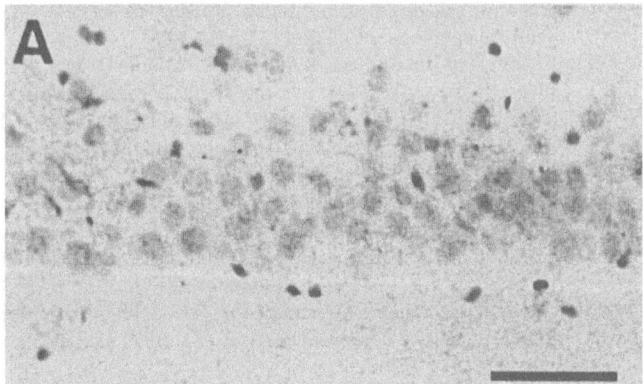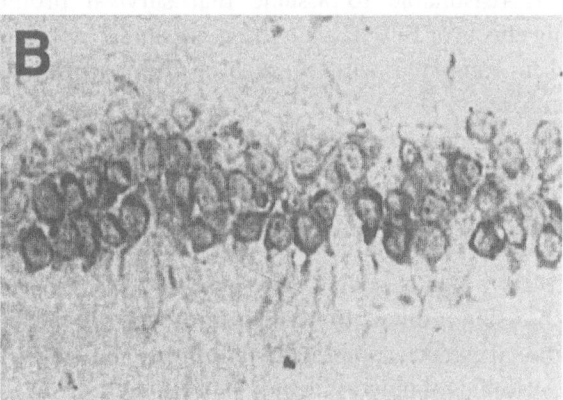

Fig. 3. Immunohistochemical detection of the heat shock protein hsp70, using the monoclonal antibody C-92F3A-5 (StressGen Biotech., Canada) and the ABC method (Vector Laboratories, Burlingame, USA). Cell nuclei were lightly counter-stained with hematoxylin. (A) Hippocampal CA1 neurons of a sham-operated control animal, surviving the sham-operation for 24 hours. (B) Hippocampal CA1 neurons of an animal surviving 30 minutes forebrain ischemia for 2 days. Note the normal morphological appearance of the cells and the pronounced cytoplasmic staining for hsp70 at 2 days after ischemia. Scale bar = 40 μm

types [6,19]. Also in support for the involvement of an embryonic death programme in CA1 neurons after ischemia is the finding that some developmentally regulated genes, such as grp78 [41] and MAP2c [27] are induced specifically in CA1 neurons after forebrain ischemia at survival times immediately preceding cell death. Also compatible with occurrence of PCD are some studies reporting a beneficial effect of infusion of the protein synthesis inhibitor cycloheximide on the survival of CA1 neurons following transient forebrain ischemia [14,31,36]. These findings implicate that the formation of a protein would be necessary for the progression of delayed neuronal death in the hippocampus, a suggestion which has also been made for PCD in SCG neurons upon NGF withdrawal [8]. However, cycloheximide infusion is known to induce hypothermia *in vivo* and it is well known that hypothermia can protect CA1 neurons following forebrain ischemia [3,21]. Therefore, although the *in vivo* effects of cycloheximide are compatible with the hypothesis of the activation of PCD, the significance of these results is debatable. Further evidence for occurrence of PCD has been provided by the demonstration of DNA fragmentation in the hippocampal CA1 sector after ischemia [11,15]. The activation of endonucleases cleaving DNA at internucleosomal sites is one of the hallmarks of most PCD paradigms, and occurs also in the SCG neurons undergoing PCD [8]. Another indication for PCD in CA1 neurons in the rat after transient forebrain ischemia is that CA1 neurons can be rescued by high doses of brain derived neurotrophic factor (BDNF) [1]. Although the precise roles of BDNF in the CNS are not completely understood it appears to be an important factor for survival of neurons within the CNS [13]. Since it is closely related to NGF, it is reasonable to assume that survival promoting effects of BDNF are exerted via similar pathways when compared with those of NGF in peripheral neurons.

In summary, several observations are indicative of a relationship of delayed neuronal death in CA1 neurons following transient forebrain ischemia and programmed cell death as it occurs in the developing central nervous system. The investigations on gene expression profiles after cerebral ischemia provide additional evidence pointing in this direction. Clearly, *in vivo* studies aiming to identify a functional role of the induced genes for delayed neuronal death following transient forebrain ischemia are needed. Since methods are now emerging which will enable us to modulate gene expression in the intact CNS, we can hope to clarify these questions in the near future. The identification of crucial components of a neuronal death programme activated by cerebral ischemia might allow us to develop therapeutic strategies to rescue cells at a very late stage after the initial ischemic insult.

References

1. Beck T, Lindholm D, Castrén E, Wree A (1994) Brain-derived neurotrophic factor protects against ischemic cell damage in rat hippocampus. J Cereb Blood Flow Metab 14: 689–692
2. Breathnach R, Matrisian LM, Gesnel MC, Staub A, Leroy P (1987) Sequences coding for part of oncogene-induced transin are highly conserved in a related rat gene. Nucleic Acids Res 15: 1139–1151
3. Busto R, Dietrich WD, Globus MYT, Valdés I, Scheinberg P, Ginsberg MD (1987) Small differences in intraischemic brain temperature critically determine the extent of ischemic neuronal damage. J Cereb Blood Flow Metab 7:729–738
4. Buttyan R, Zakeri Z, Lockshin R, Wolgemuth D (1988) Cascade induction of c-fos, c-myc, and heat shock 70K transcripts during regression of the rat ventral prostate gland. Mol Endocrinol 2: 650–657
5. Buttyan R, Olsson CA, Pintar J, Chang C, Bandyk M, Ng PY, Sawczuk IS (1989) Induction of the TRPM-2 gene in cells undergoing programmed death. Mol Cell Biol 9: 3473–3481
6. Clarke AR, Purdie CA, Harrison DJ, Morris RG, Bird CC, Hooper ML, Wyllie AH (1993) Thymocyte apoptosis induced by p53-dependent and independent pathways. Nature 362: 849–852
7. Colotta F, Polentarutti N, Sironi M, Mantovani A (1992) Expression and involvement of c-fos and c-jun protooncogenes in programmed cell death induced by growth factor deprivation in lymphoid cell lines. J Biol Chem 267: 18278–18283
8. Deckwerth TL, Johnson EM Jr (1993) Temporal analysis of events associated with programmed cell death (apoptosis) of sympathetic neurons deprived of nerve growth factor. J Cell Biol 123: 1207–1222
9. Dragunow M, Young D, Hughes P, MacGibbon G, Lawlor P, Singleton K, Sirimanne E, Beilharz E, Gluckman P (1993) Is c-jun involved in nerve cell death following status epilepticus and hypoxic-ischemic brain injury? Mol Brain Res 18: 347–352
10. Estus S, Zaks WJ, Freeman RS, Gruda M, Bravo R, Johnson EM Jr (1994) Altered gene expression in neurons during programmed cell death: identification of c-jun as necessary for neuronal apoptosis. J Cell Biol 127: 1717–1727
11. Ferrer I, Tortosa A, Macaya A, Sierra A, Moreno D, Munell F, Blanco R, Squier W (1994) Evidence of nuclear DNA fragmentation following hypoxia-ischemia in the infant rat brain, and transient forebrain ischemia in the adult gerbil. Brain Pathol 4: 115–122
12. Freeman RS, Estus S, Johnson EM Jr (1994) Analysis of cell-cycle-related gene expression in postmitotic neurons: selective induction of Cyclin D1 during programmed cell death. Neuron 12: 343–355
13. Ghosh A, Carnahan J, Greenberg ME (1994) Requirement for BDNF in activity-dependent survival of cortical neurons. Science 263: 1618–1623
14. Goto K, Ishige A, Sekiguchi K, Iizuka S, Sugimoto A, Yuzurihara M, Aburada M, Hosoya E, Kogure K (1990) Effects of cycloheximide on delayed neuronal death in rat hippocampus. Brain Res 534: 299–302
15. Heron A, Pollard H, Dessi F, Moreau J, Lasbennes F, Ben-Ari Y, Charriaut-Marlangue CR (1993) Regional variability in

DNA fragmentation after global ischemia evidenced by combined histological and gel electrophoresis observations in rat brain. J Neurochem 61: 1973–1976

16. Hossmann KA (1993) Disturbances of cerebral protein synthesis and ischemic cell death. Progr Brain Res 96: 161–177

17. Jörgensen MB, Deckert J, Wright DC, Gehlert DR (1989) Delayed c-fos protooncogene expression in the rat hippocampus induced by transient global cerebral ischemia: an in situ hybridization study. Brain Res 484: 393–398

18. Kiessling M, Stumm G, Xie Y, Herdegen T, Aguzzi A, Bravo R, Gass P (1993) Differential expression and translation of immediate early genes in the gerbil hippocampus after transient forebrain ischemia. J Cereb Blood Flow Metab 13: 914–924

19. Lowe SW, Schmitt EM, Smith SW, Osborne BA, Jacks T (1993) p53 is required for radiation-induced apoptosis in mouse thymocytes. Nature 362: 847–849

20. Martin SJ, Green DR, Cotter TG (1994) Dicing with death: dissecting the components of the apoptosis machinery. Trends Biochem Sci January: 26–30

21. Minamisawa H, Nordström CH, Smith ML, Siesjö BK (1990) The influence of mild body and brain hypothermia on ischemic brain damage. J Cereb Blood Flow Metab 10: 817–822

22. Neumann-Haefelin T, Wiessner C, Vogel P, Back T, Hossmann KA (1994) Differential expression of the immediate early genes c-fos, c-jun, junB, and NGFI-B in the rat following transient forebrain ischemia. J Cereb Blood Flow Metab 14: 206–216

23. Nowak TS Jr, Osborne OC, Suga S (1993) Stress protein and proto-oncogene expression as indicators of neuronal pathophysiology after ischemia. In: Kogure K, Hossmann KA, Siesjö BK (eds) Progress in brain research, Vol 96: neurobiology of ischemic brain damage. Elsevier, Amsterdam, pp 195–208

24. Oppenheim RW (1991) Cell death during development of the nervous system. Annu Rev Neurosci 14: 453–501

25. Pulsinelli WA, Brierley JB (1979) A new model of bilateral hemispheric ischemia in the unanesthetized rat. Stroke 10: 267–272

26. Safi E, Manning F, Fiskum G, Rosenthal R, Patierno S (1992) Stress response genes and a gene associated with apoptosis are induced in a canine model of global cerebral ischemia and reperfusion. FASEB J 6(4): A1063

27. Saito N, Kawai K, Nowak TS Jr (1995) Reexpression of developmentally regulated MAP2c mRNA after ischemia: colocalization with hsp72 mRNA in vulnerable neurons. J Cereb Blood Flow Metab 15: 205–215

28. Schmidt-Kastner R, Paschen W, Grosse Ophoff B, Hossmann KA (1989) A modified four-vessel occlusion model for inducing incomplete ischemia in rats. Stroke 20: 938–946

29. Siesjö BK (1988) Historical overview. Calcium, ischemia, and death of brain cells. Ann NY Acad Sci 522: 638–661

30. Siesjö BK, Lundgren J, Pahlmark K (1990) The role of free radicals in ischemic brain damage: a hypothesis. In: Krieglstein J, Oberpichler H (eds) Pharmacology of cerebral ischemia. Wissenschaftliche Verlagsgesellschaft, Stuttgart, pp 319–323

31. Shigeno T, Yamasaki Y, Kato G, Kusaka K, Mima T, Takakura K, Graham DI, Furukawa S (1990) Reduction of delayed neuronal death by inhibition of protein synthesis. Neurosci Lett 120: 117–119

32. Smeyne RJ, Vendrell M, Hayward M, Baker SJ, Miao GG, Schilling K, Robertson LM, Curran T, Morgan JI (1993) Continuous c-fos expression precedes programmed cell death in vivo. Nature 363: 166–169

33. Stubberöd P, Tomasevic G, Kamme F, Wieloch T (1994) Expression of the tumor supressor p53 and WAF1 in the postischemic rat hippocampus. Abstr Soc Neurosci 20: 616

34. Sun H, Charles CH, Lau LF, Tonks NK (1993) MKP-1 (3CH134), an immediate early gene product, is a dual specificity phosphatase that dephosphorylates MAP kinase in vivo. Cell 75: 487–493

35. Tomioka C, Nishioka K, Kogure K (1993) A comparison of induced heat-shock protein in neurons destined to survive and those destined to die after transient ischemia in rats. Brain Res 612: 216–220

36. Tortosa A, Rivera R, Ferrer I (1994) Dose-related effects of cycloheximide on delayed neuronal death in the hippocampus after bilateral transitory forebrain ischemia. J Neurol Sci 12: 10–17

37. Wessel TC, Joh TH, Volpe BT (1991) In situ hybridization analysis of c-fos and c-jun expression in the rat brain following transient ischemia. Brain Res 567: 231–240

38. Widmann R, Miyazawa T, Hossmann KA (1993) Protective effect of hypothermia on hippocampal injury after 30 minutes of forebrain ischemia in rats is mediated by postischemic recovery of protein synthesis. J Neurochem 61: 200–209

39. Wiessner C, Back T, Bonnekoh P, Kohno K, Gehrmann J, Hossmann KA (1993a) Sulfated glycoprotein-2 mRNA in the rat brain following transient forebrain ischemia. Mol Brain Res 20: 345–352

40. Wiessner C, Gehrmann J, Lindholm D, Töpper R, Kreutzberg GW, Hossmann KA (1993b) Expression of transforming growth factor-ß1 and interleukin-1ß mRNA in rat brain following transient forebrain ischemia. Acta Neuropathol: 86, 439–446

41. Wiessner C, Neumann-Haefelin T, Brink I, Vogel P, Hossmann KA (1994) The genomic response of the rat brain following cerebral ischemia: generalized versus cell and region specific responses. In: Krieglstein J, Oberpichler-Schwenk H (eds) Pharmacology of cerebral ischemia. Wissenschaftliche Verlagsgesellschaft, Stuttgart, pp 455–465

42. Wiessner C (1995) The dual specificity phosphatase PAC-1 is transcriptionally induced in the rat brain following transient forebrain ischemia. Mol Brain Res 28: 353–356

43. Wiessner C, Neumann-Haefelin T, Vogel P, Back T, Hossmann KA (1995) Transient forebrain ischemia induces an immediate-early gene encoding the mitogen-activated protein kinase phosphatase 3CH134 in the adult rat brain. Neuroscience 64: 959–966

44. Zoli M, Ferraguti F, Zini I, Bettuzzi S, Agnati LF (1993) Increases in sulphated glycoprotein-2 mRNA levels in the rat brain after transient forebrain ischemia or partial mesodiencephalic hemitranssection. Mol Brain Res 18: 163–177

Correspondence: Christoph Wiessner, M.D., Max-Planck-Institute for Neurological Research, Department for Experimental Neurology, Gleueler Str. 50, D-50931 Köln, Federal Republic of Germany.

Acta Neurochir (1996) [Suppl] 66: 8–14

Molecular Mechanisms of Acidosis-Mediated Damage

B. K. Siesjö[1], **K. I. Katsura**[1], **T. Kristián**[1], **P.-A. Li**[1], and **P. Siesjö**[2]

[1]Laboratory for Experimental Brain Research and [2]Department of Tumor Immunology, University of Lund, Lund, Sweden

Summary

The present article is concerned with mechanisms which are responsible for the exaggerated brain damage observed in hyperglycemic animals subjected to transient global or forebrain ischemia. Since hyperglycemia enchances the production of lactate plus H^+ during ischemia, it seems likely that aggravation of damage is due to exaggerated intra- and extracellular acidosis. This contention is supported by results showing a detrimental effect of extreme hypercapnia in normoglycemic rats subjected to transient ischemia or to hypoglycemic coma.

Enhanced acidosis may exaggerate ischemic damage by one of three mechanisms: (i) by accelerating free radical production via H^+-dependent reactions, some of which are catalyzed by iron released from protein bindings by a lowering of pH, (ii) by perturbing the intracellular signal transduction pathway, leading to changes in gene expression or protein synthesis, or (iii) by activating endonucleases which cause DNA fragmentation .

While activation of endonucleases must affect the nucleus, the targets of free radical attack are not known. Microvessels are considered likely targets of such attack in sustained ischemia and in trauma; however, enhanced acidosis is not known to aggravate microvascular dysfunction, or to induce inflammatory responses at the endothelial-blood interface. A more likely target is the mitochondrion. Thus, if the ischemia is of long duration (30 min) hyperglycemia triggers rapidly developing mitochondrial failure. It is speculated that this is because free radicals damage components of the respiratory chain, leading to a secondary deterioration of oxidative phosphorylation.

Keywords: Ischemia; acidosis; free radicals; mitochondria.

Introduction

When sufficiently severe and sufficiently prolonged, ischemia leads to irreversible cell damage, initially affecting neurons (yielding selective neuronal damage), then glial cells and vascular tissues (resulting in infarction). The mechanisms involved are probably multifactorial, involving calcium "overload", acidosis, free radicals, and inflammatory reactions [54]. In the transition from selective neuronal necrosis to infarction, the duration of ischemia plays an important role. However, evidence now exists that exaggerated tissue acidosis is an important factor in this transition; in fact, a condition like hypoglycemic coma (which does not lead to acidosis) can lead to massive neuronal necrosis, yet sparing glial cells and vascular tissue [47,52].

The evidence linking acidosis to enhanced tissue damage is mainly based on experiments showing that preischemic hyperglycemia aggravates the damage caused by a transient period of ischemia, and that this is due to exaggerated intra- and extracellular acidosis [39,47,51,52]. This contention is not accepted by all since there are data showing that acidosis ameliorates, rather than aggravates damage to neurons caused by glutamate and anoxia [60]. However, the results quoted in support of this alternative hypothesis are almost entirely based on *in vitro* experiments; besides, additional *in vivo* results strongly support the notion that acidosis aggravates ischemic damage [25]. We will thus base our discussion on the postulate that, *in vivo*, acidosis aggravates ischemic brain damage (see also [30], for data on hypoglycemic coma) and consider possible mechanisms of such acidosis-mediated cell damage.

1. Acidosis and Cell Death: Putative Mechanisms

Acidosis affects a multitude of metabolic processes, and modulates membrane processes such as dissipative and active transport of ions [5,36]. In general, acidosis depresses metabolic reactions and ion flux. Some of its effects on membrane functions, e.g. on calcium conductances, may even have a protective value, as observed when extracellular acidosis protects neurons *in vitro* from the harmful effects of glutamate exposure, or anoxia [14,59]. However, our present problem is a different one since it concerns the effects of exaggerated acidosis on an energy-compromised tissue, in which the cellular phosphorylation potential is reduced and, as a consequence, a cascade of potentially harmful metabolic reactions has been elicited [47,51]. It is also important to emphasize that the harmful effect of acidosis is exerted at the time of the insult, i.e. during and/or just after the transient ischemia or hypoglycemia [30]. Thus, since the adverse effects of hyperglycemia (and acidosis) are manifested after a delay of hours or days [46,63,64], we are confronted with the problem: how can exaggerated acidosis during and/or just after the insult give rise to damage which "matures" over many hours? The following mechanisms can be envisaged.

a. Acidosis Causes Enhanced Production of Free Radicals

As discussed elsewhere [47,48,51,52] a reduction of pH enhances free radical production in brain homogenates. This effect was observed three decades ago [1] and was subsequently described and analyzed in more detail [44,50]. In these *in vitro* experiments, free radical production was triggered by addition of iron, and iron chelators were found to blunt the reaction [44]. This makes it likely that acidosis acts by releasing iron from its binding to transferrin-like proteins according to the reaction [48,50,52].

$$Fe^{3+} + HCO_3^- + H_3TFn \leftrightarrow FeTFnHCO_3^- + 3H^+ \quad (1)$$

However, free radicals can be formed in reactions which are not necessarily triggered by iron. One of these is the conversion of the superoxide anion ($\cdot O_2^-$) to its acid form ($H \cdot O_2$) according to the reaction

$$\cdot O_2^- + H^+ \leftrightarrow H \cdot O_2, pK_a = 4.7 \quad (2)$$

Another set of reactions of potential importance for H^+-induced enhancement of free radical formation comprises those in which generation of $\cdot O_2^-$ and nitric oxide ($N \cdot O$), and their interaction, leads to the production of $\cdot OH(4)$. We can write the following reactions

$$N \cdot O + \cdot O_2^- \rightarrow ONOO^- \quad (3)$$

$$ONOO^- + H^+ \rightarrow ONOOH \quad (4)$$

$$ONOOH \rightarrow \cdot OH + NO_2 \quad (5)$$

In summary interaction between $N \cdot O$ and $\cdot O_2^-$ yields peroxynitrite ($ONOO^-$) which, after protonation, decomposes with the production of $\cdot OH$. However, it still remains to be shown if acidosis has any effect on this set of reactions.

Although free radicals are likely mediators of acidosis-mediated ischemic brain damage it has been difficult to unequivocally prove their importance. For example, although dimethylthiourea (DMTU), a hydroxyl radical scavenger, ameliorates damage due to transient ischemia in normoglycemic animals [41], the effect of DMTU in hyperglycemic animals is limited [32]. It remains to be shown, therefore, to what extent free radicals contribute to the exaggeration of ischemic damage in hyperglycemic animals, and what the target of attack is.

b. Acidosis Gives Rise to DNA Damage, or Perturbs Signal Transduction

Experiments conducted 15 years ago showed that preischemic hyperglycemia leads to condensation of nuclear chromatin in neurons [23]. The mechanisms behind this morphological change have never been clarified, nor has it been shown whether the alterations observed are pathogenetically important. It is not unlikely, though, that hyperglycemia acts by further perturbing the signal transduction chain or by enhancing DNA fragmentation. To take one example, hyperglycemia exaggerates the sustained metabolic depression which is observed after a transient period of ischemia ([28], see Fig. 1). Furthermore, Combs *et al.*

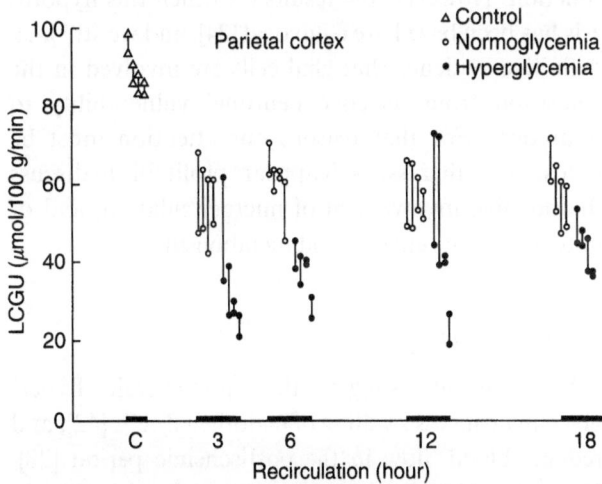

Fig. 1. The influence of preischemic hyperglycemia on postischemic local cerebral glucose utilization (LCGU), following 10 min of forebrain ischemia in rats. LCGU was measured with the autoradiographic deoxyglucose technique of Sokoloff *et al.* [56]. Values from the two hemispheres were joined by a vertical line. Data from Kozuka *et al.* [28]

(1992) found that hyperglycemia blunted or aborted the expression of mRNA for c-fos, normally caused by ischemia and reperfusion [6]. Finally, data obtained on non-neuronal cells suggest the presence of a pH-dependent endonuclease (DNAse II) which is activated when pH falls below 6.5 [2,3]. Clearly, it seems highly justified to focus additional attention on the effect of acidosis on control of metabolic cascades, and on DNA-linked events.

c. Acidosis Aggravates Cell Edema

There are theoretical reasons why a lowered pH should enhance the formation of cell edema [48,52] and, *in vitro*, acidosis enhances cell swelling [26]. However , postischemic edema is not a conspicuous feature in hyperglycemic animals recovering from an ischemic transient, and massive edema is not observed until 18–24 hours later [64]. The secondary edema is likely a result of a different kind of pathology, one affecting the integrity of the blood-brain barrier (BBB), or mitochondrial respiratory functions.

2. Targets of Attack

It is still an unresolved question whether hyperglycemia (and exaggerated acidosis) aggravates tissue damage by acting on neurons, on glial cells, or on microvessels. A glial origin was postulated by Plum [43], and some supportive data were published by Kraig *et al.* [29] who postulated that hyperglycemia leads to infarction by disrupting glial metabolism and function. However, the results on which this hypothesis has been based are equivocal [24], and we still lack definitive evidence that glial cells are involved in the transition from selective neuronal vulnerability to infarction. For that reason, our attention must be focused on other issues. It appears profitable to discuss the possible involvement of microcirculation, and of mitochondrial function and metabolism.

a. Microvessels

Some results suggest that preischemic hyperglycemia triggers swelling of endothelial cells [42], and reduces blood flow in the postischemic period [28]. However, it has not been shown that the reduction in blood flow is pathophysiologically important, nor has it been shown that hyperglycemia and/or acidosis cause microcirculatory problems predisposing to infarction. The only solid evidence in favor of the hypothesis that acidosis causes microcirculatory problems is that presented by Nedergaard and Diemer (1987), who found that the nature of the tissue lesions following MCA occlusion was a function of the preischemic plasma glucose concentration [39]. Essentially, preischemic hypoglycemia was found to predispose to selective neuronal necrosis, and hyperglycemia to infarction, with destruction of all cellular elements, including microvessels. In theory, such results could reflect the effect of acidosis on the production of free radicals during and immediately following ischemia, notably that caused by delocalization of iron from its binding to proteins. It has been emphasized that microvessels constitute a likely target of free radical attack [27,49]. Evidence in support of this contention is the high concentration of xanthine oxidase in cultured endothelial cells [58], and the generation of $\cdot$OH in microvessel preparations challenged with anoxia and reoxygenation [16]. It remains to be shown, though, whether the pan-necrosis reflects "specific" damage to the microcirculation or the indiscriminate attack on all cellar elements, with secondary effects on microvessels. It also remains to be shown whether or not acidosis has any effect on adhesion of polymorphonuclear leukocytes to endothelial cells, or on inflammatory responses at the blood-endothelial cell interface [7,13,17]. In general, the effect of acidosis on microcirculatory events remains unclarified.

b. Mitochondria

Mitochondria represent a likely target of free radical damage. First, electron transport is normally associated with a small electron leak, i.e. with reactions in which univalent reduction of O_2 yields partially reduced oxygen species (ROS), notably $\cdot O_2^-$ and H_2O_2, and this electron leak can increase in pathological states [18]. Second, when mitochondria are exposed to a free radical generating system *in vitro*, state 3 respiration is affected in much the same way as it is after ischemia with recirculation [19,20].

(i) Ischemia and early recirculation. Mitochondria isolated from ischemic tissue show a reduction in state 3 respiration, which is usually most marked when malate plus glutamate are used as substrates [15,21, 45,55]. However, following reasonably brief periods of ischemia (e.g. 15 min) respiratory functions of isolated mitochondria return to normal. This is illustrated in the left panel of Fig. 2 which gives data for hyperglycemic subjects [22]. A similar return of mitochondrial function was reported by Sims and Pulsinelli

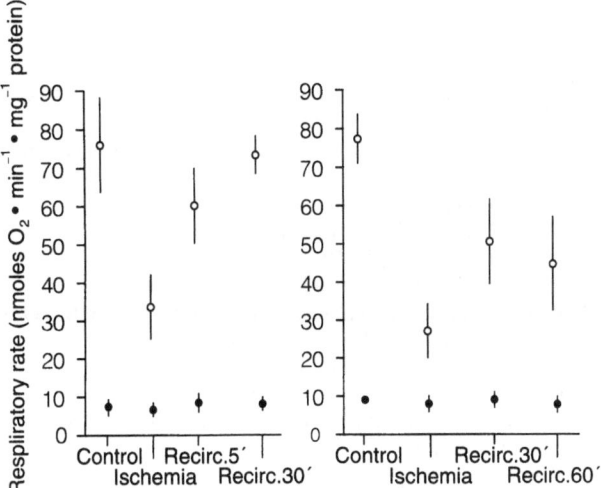

Fig. 2. Integrated respiratory functions of mitochondria isolated from rat brains during ischemia and recirculation. Unfilled circles denote state 3 and filled circles state 4 respiration with glutamate plus malate as substrates. Values are means with 95 % confidence limits indicated. All animals were hyperglycemic. The left panel demonstrates recovery of respiratory functions after 15 min of ischemia, the right one lack of recovery after 30 min of ischemia. Reconstructed from Hillered et al. [21,22]

(1987) who employed 30 min of forebrain ischemia in starved rats [55].

The inhibition of state 3 respiration, and the general depression of mitochondrial functions, may reflect hydrolysis of phospholipids by activation of lipolytic enzymes (primarily phospholipase A$_2$), and accumulation of polyunsaturated free fatty acids (FFAs). This contention is supported by results demonstrating breakdown of mitochondrial phospholipids and accumulation of FFAs following ischemia [31,33,37,56a, 57]; in addition, when isolated mitochondria are exposed to polyenoic FFA, they are affected in much the same way as in ischemia [19,31,57]. It is not known how arachidonic acid exerts its adverse effects but, apart from acting as an ionophore, it could serve as a substrate in reactions leading to the production of free radicals. That free radicals are formed is suggested by the accumulation of malondialdehyde (MDA) in mitochondria from ischemic animals [34,56a]. It has also been clearly shown that free radicals affect mitochondrial respiration in much the same way as ischemia does. For example, mitochondria exposed to increased concentrations of Ca^{2+}, Na^+, and the peroxide tertbutyl hydroperoxide (tBOOH), i.e. to conditions prevailing during ischemia in vivo, showed dramatic deterioration of respiratory functions [61]. The effects required that calcium entered mitochondria, and the adverse effects were thus calcium-mediated, possibly reflecting calcium activation of PLA$_2$. A later study

showed that mitochondria from cerebral tissues, exposed to 2.5 µm Ca^{2+}, 14 mM Na^+, and elevated ADP under normoxic conditions, generate the extremely toxic ·OH species, plus ascorbyl and other carbon-centered radicals [12]. The conditions were found to impair the function of NADH-CoQ reductase (Complex I), explaining the preferential reduction of state 3 respiration, with pyruvate or glutamate plus malate as substrates. However, the site of attack in ischemic tissue is a controversial issue, particularly since Zaiden and Sims (1993) reported a clear reduction of the activity of the pyruvate dehydrogenase complex ([65], however, see also [31]).

As shown in Fig. 2 (right panel), prolongation of the ischemia to 30 min gave rise to a sustained depression of state 3 respiration. Similar results have been reported by others [31,45,56a]. On analysis, though, animals subjected to 30 min of ischemia in the starved state showed recovery of mitochondrial respiratory functions upon reperfusion, while those with documented hyperglycemia did not [21,31,45].

(ii) Ischemia and late recirculation. Resumption of normal mitochondrial respiratory functions after brief to intermediate periods of ischemia occurs pari passu with recovery of the bioenergetic state (e.g. [22]). In areas destined to incur delayed neuronal necrosis cell death is preceded by a scond phase of mitochondrial dysfunction, involving a decrease in ADP- and uncoupler-stimulated respiration [55]. A likely scenario is that gradual accumulation of calcium in cells with a perturbed plasma membrane function leads to mitochondrial calcium "overload" [8,9,11]. Conceivably, this leads to calcium mediated free radical generation, e.g. via activation of phospholipase A$_2$ [12,35,61]. Support for this contention was recently reported by Zaiden and Sims (1994) who noted a secondary increase in mitochondrial calcium content in caudoputamen 3–6 hrs after the start of recirculation, following 30 min of forebrain ischemia [66]. This calcium accumulation seemed to precede deterioration of cellular energy state, making it likely that the latter was due to calcium overload of mitochondria, as predicted by Deshpande et al. [8]. However, it is not known whether mitochondrial dysfunction is the cause or the effect of net calcium uptake by the mitochondria.

(iii) The effect of hyperglycemia / exaggerated acidosis. The question arises whether exaggerated acidosis enhances damage to mitochondria, and whether any adverse effect is related to increased production of free radicals. The results of Rehncrona et al. (1979) showed that a 30 min period of complete and incomplete ischemia reduced state 3 respiration equally; however, recirculation failed to normalize mitochondrial respiratory activity in animals with incomplete ischemia, who developed excessive lactic acidosis [45]. Subsequent experiments from our laboratory confirmed these findings [21], and additional results were reported by Wagner *et al.* [62,63]. These authors, employing a model with transient anoxia in normo- and hyperglycemic cats, sampled tissue for mitochondrial isolation in symptom-free animals, as well as in those developing symptoms of neurological complications. The latter yielded mitochondria with markedly inhibited substrate-, ADP-, and uncoupler stimulated respiratory rates. Interestingly, in these mitochondria the activities of cytochrome oxidase and cytochrome a-a$_3$ were reduced, suggesting an aberrant type of mitochondrial dysfunction.

The results reported in hyperglycemic animals subjected to 15 min of forebrain ischemia [22] demonstrate an initial recovery of mitochondrial functions in a model which subsequently leads to extensive tissue damage and irreversible seizures, while those of Wagner *et al.* [62,63] reveal mitochondrial failure in hyperglycemic subjects who develop postischemic symptoms. It is tempting to conclude that delayed mitochondrial failure is responsible for the rapid maturation of damage, but it is not known what is the cause and the effect, nor is it known if calcium overload of the mitochondria is involved. However, evidence exists that the combination of a rise in Ca^{2+}, ATP hydrolysis, and oxidative stress can create a leak for H$^+$ across the inner mitochondrial membrane, yielding mitochondrial dysfunction [10].

In conclusion, evidence exists that hyperglycemia (and enhanced intraischemic acidosis) prevents recovery of mitochondrial functions after long periods of ischemia. This suggests that acidosis triggers rapidly developing mitochondrial failure, possibly by mechanisms related to enhanced formation of free radicals. The situation is different after shorter periods of ischemia, since mitochondrial function is then resumed in tissues which are destined to incur massive damage. It seems highly justified to explore whether this damage is due to delayed, free radical-mediated mitochondrial injury.

Acknowledgements

Work discussed in this article was supported by grants from the Swedish Medical Research Council, and from the U.S. Public Health Service via the NINDS.

References

1. Barber AA, Bernheim F (1967) Lipid peroxidation: its measurement, occurrence and significance in animal tissues. Adv Gerontol Res 2: 355–403
2. Barry MA, Eastman A (1993) Identification of deoxyribonuclease II as an endonuclease involved in apoptosis. Arch Biochem Biophys 300: 440–450
3. Barry MA, Reynolds JE, Eastman A (1993) Etoposide-induced apoptosis in human HL-60 cells is associated with intracellular acidification. Canc Res 53: 2349–2357
4. Beckman J, Beckman T, Chen J, Marshall P, Freeman B (1990) Apparent hydroxyl radical production by peroxynitrite: implications for endothelial injury from nitric oxide and superoxide. Proc Natl Acad Sci 87: 1620–1624
5. Busa WB, Nuccitelli R (1984) Metabolic regulation via intracellular pH. Am J Physiol 246: R409–438
6. Combs DJ, Dempsey RJ, Donaldson D, Kindy MS (1992) Hyperglycemia suppresses c-fos mRNA expression following transient cerebral ischemia in Gerbils. J Cereb Blood Flow Metab 12: 169–172
7. del Zoppo GJ (1994) Microvascular changes during cerebral ischemia and reperfusion. Cerebrovasc Brain Metab Rev 6: 47–96
8. Deshpande JK, Siesjö BK, Wieloch T (1987) Calcium accumulation and neuronal damage in the rat hippocampus following cerebral ischemia. J Cereb Blood Flow Metab 7: 89–95
9. Dienel GA (1984) Regional accumulation of calcium in postischemic rat brain. J Neurochem 43: 913–925
10. Duchen M, McGuinness O, Brown L, Crompton M (1994) On the involvement of a cyclosporin A sensitive mitochondrial pore in myocardial reperfusion injury. Cardiovasc Res 27: 1790–1794
11. Dux E, Mies G, Hossmann K-A, Siklos L (1987) Calcium in the mitochondria following brief ischemia of gerbil brain. Neurosci Lett 78: 295–300
12. Dykens JA (1994) Isolated cerebral and cerebellar mitochondria produce free radicals when exposed to elevated Ca^{2+} and Na$^+$: implications for neurodegeneration. J Neurochem 63: 584–591
13. Feuerstein GZ, Liu T, Barone FC (1994) Cytokines, inflammation, and brain injury: role of tumor necrosis factor-a. Cerebrovasc Brain Metab Rev 6: 341–360
14. Giffard RG, Monyer H, Christine CW, Choi DW (1990) Acidosis reduces NMDA receptor activation, glutamate neurotoxicity, and oxygen-glucose deprivation neuronal injury in cortical cultures. Brain Res 506: 339–342
15. Ginsberg MD, Mela L, Wrobel-Kuhl K, Reivich M (1977) Mitochondrial metabolism following bilateral cerebral ischemia in the gerbil. Ann Neurol 1: 519–527
16. Grammas P, Liu G-J, Wood K, Floyd RA (1993) Anoxia/reoxygenation induces hydroxyl free radical formation in brain microvessels. Free Radic Biol Med 14: 553–557
17. Hallenbeck J (1996) Inflammatory reactions at the blood-endothelial interface in acute stroke. In: Siesjö B, Wieloch T (eds) Advances in neurology, vol 69. New York, Raven, in press
18. Halliwell B, Gutteridge JMC (1985) Oxygen radicals and the nervous system. Trends Neurosci 8: 22–26
19. Hillered L, Chan PH (1988) Role of arachidonic acid and other free fatty acids in mitochondrial dysfunction in brain ischemia. J Neurosci Res 20: 451–456
20. Hillered L, Ernster L (1983) Respiratory activity of isolated rat brain mitochondria following in vitro exposure to oxygen radicals. J Cereb Blood Flow Metab 3: 207–214
21. Hillered L, Ernster L, Siesjö BK (1984) Influence of in vitro lactic acidosis and hypercapnia on respiratory activity of isolated rat brain mitochondria. J Cereb Blood Flow Metab 4: 430–437

22. Hillered L, Smith M-L, Siesjö BK (1985) Lactic acidosis and recovery of mitochondrial function following forebrain ischemia in the rat. J Cereb Blood Flow Metab 5: 259–266

23. Kalimo H, Rechncrona S, Söderfeldt B, Olsson Y, Siesjö BK (1981) Brain lactic acidosis and ischemic cell damage: 2. Histopathology. J Cereb Blood Flow Metab 1: 313–327

24. Katsura K, Asplund B, Ekholm A, Siesjö BK (1992) Extra- and intracellular pH in the brain during ischemia, related to tissue lactate content in normo- and hypercapnic rats. Eur J Neurosci 4: 166–176

25. Katsura K, Kristián T, Smith M-L, Siesjö BK (1994) Acidosis induced by hypercapnia exaggerates ischemic brain damage. J Cereb Blood Flow Metab 14: 243–250

26. Kempski O, Staub F, Jansen M, Schödel F, Baethmann A (1988) Glial swelling during extracellular acidosis in vitro. Stroke 19: 385–392

27. Kontos HA (1989) Oxygen radicals in cerebral ischemia. In: Ginsberg M, Dietrich W (eds) Cerebrovascular diseases. Raven, New York, p 365–371

28. Kozuka M, Smith M-L, Siesjö BK (1989) Preischemic hyperglycemia enhances postischemic depression of cerebral metabolic rate. J Cereb Blood Flow Metab 9: 478–490

29. Kraig RP, Pulsinelli WA, Plum F (1986) Carbonic acid buffer changes during complete brain ischemia. Am J Physiol 250: R348–R357

30. Kristián T, Gidö G, Siesjö BK (1995) The influence of acidosis on hypoglycemic brain damage. J Cereb Blood Flow Metab 15: 78–87

31. Kurup CKR, Kumaroo KK, Dutka AJ (1990) Influence of cerebral ischemia and post-ischemic perfusion on mitochondrial oxidative phosphorylation. J Bioenerg Biomemb 22: 61–80

32. Lundgren J, Smith M-L, Siesjö BK (1992) Effects of dimethylthiourea on ischemic brain damage in hyperglycemic rats. J Neurol Sci 113: 187–197

33. Majewska MD, Lazarewicz J, Strosznajder J (1977) Catabolism of mitochondrial membrane phospholipids in conditions of ischemia and barbiturate anesthesia. Bull Acad Pol Sci 25: 125–131

34. Majewska MD, Strosznajder J, Lazarewicz J (1978) Effect of ischemic anoxia and barbiturate anesthesia on free radical oxidation of mitochondrial phospholipids. Brain Res 158: 423–434

35. Malis CD, Bonventre JV (1986) Mechanism of calcium potentiation of oxygen free radical injury to renal mitochondria. J Biol Chem 261: 14201–14208

36. Moody W (1984) Effects of intracellular H^+ on the electrical properties of excitable cells. Ann Rev Neurosci 7: 257–278

37. Nakahara I, Kikuchi H, Taki W, Nishi S, Kito M, Yonekawa Y, Goto Y, Ogata N (1991) Degradation of mitochondrial phospholipids during experimental cerebral ischemia in rats. J Neurochem 57: 839–844

38. Nedergaard M (1987) Transient focal ischemia in hyperglycemic rats is associated with increased cerebral infarction. Brain Res 408: 79–85

39. Nedergaard M, Diemer N (1987) Focal ischemia of the rat brain, with special reference to the influence of plasma glucose concentration. Acta Neuropathol (Berl) 73: 131–137

40. Nedergaard, M. (1988) Mechanisms of brain damage in focal cerebral ischemia. Acta Neurol Scand 77: 1–23

41. Pahlmark K, Folbergrová J, Smith M-L, Siesjö BK (1993) Effects of dimethylthiourea on selective neuronal vulnerability in forebrain ischemia in rats. Stroke 24: 731–737

42. Paljärvi L, Rehncrona S, Söderfeldt B, Olsson Y, Kalimo H (1983) Brain lactic acidosis and ischemic cell damage: quantitative ultrastructural changes in capillaries of rat cerebral cortex. Acta Neuropathol (Berl) 60: 232–240

43. Plum F (1983) What causes infarction in ischemic brain ?: The Robert Wartenberg lecture. Neurology 33: 222–233

44. Rehncrona S, Hauge HN, Siesjö BK (1989) Enhancement of iron-catalyzed free radical formation by acidosis in brain homogenates: difference in effect by lactic acid and CO_2. J Cereb Blood Flow Metab 9: 65–70

45. Rehncrona S, Mela L, Siesjö BK (1979) Recovery of brain mitochondrial function in the rat after complete and incomplete cerebral ischemia. Stroke 10: 437–446

46. Siemkowicz E, Hansen A (1978) Clinical restitution following cerebral ischemia in hypo-, normo-, and hyperglycemic rats. Acta Neurol Scand 58: 1–8

47. Siesjö BK (1988) Mechanisms of ischemic brain damage. Crit Care Med 16: 954–963

48. Siesjö BK, (1985) Acid-base homeostasis in the brain: physiology, chemistry, and neurochemical pathology. In: Hossmann K-A, Kogure K, Siesjö BK, Welsh F(eds) Progress in brain research. Elsevier, Amsterdam, pp 121–154

49. Siesjö BK, Agardh C-D, Bengtsson F (1989) Free radicals and brain damage. Cerebrovasc Brain Metab Rev 1: 165–211

50. Siesjö BK, Bendek G, Koide T, Westerberg E, Wieloch T (1985) Influence of acidosis on lipid peroxidation in brain tissues in vitro. J Cereb Blood Flow Metab 5: 253–258

51. Siesjö B, Katsura K, Kristián T (1996) Acidosis-retated damage. In: Siesjö B, Wieloch T (eds) Advances in neurology, vol 69. Raven, New York, in press

52. Siesjö BK, Katsura K, Mellergård P, Ekholm A, Lundgren J, Smith M-L, (1993) Acidosis-related brain damage: neurobiology of ischemic brain damage. In: Kogure K, Hossmann K-A, Siesjö BK. Progress in brain research. Elsevier, Amsterdam, pp 23–48

53. Siesjö BK, Wieloch T (1985) Molecular mechanisms of ischemic brain damage: Ca^{2+}-related events. In: Plum F, Pulsinelli W (eds) Cerebrovascular diseases. Raven, New York, pp 187–200

54. Siesjö BK, Wieloch T (eds) (1996) Advances in neurology, vol 69. Raven, New York, in press

55. Sims NR, Pulsinelli WA (1987) Altered mitochondrial respiration in selectively vulnerable brain subregions following transient forebrain ischemia in the rat. J Neurochem 49: 1367–1374

56. Sokoloff L, Reivich M, Kennedy C, Des Rosiers MH, Patlak CS, Pettigrew KD, Sakurada O, Shinohara M (1977) The [14C] deoxyglucose method for the measurement of local cerebral glucose utilization: theory, procedure, and normal values in the conscious and anesthetized albino rat. J Neurochem 28: 897–916

56a. Sun D, Gilboe DD (1994) Ischemia-induced changes in cerebral mitochondrial free fatty acids, phospholipids, and respiration in the rat. J Neurochem 62: 1921–1928

57. Takeuchi Y, Morii H, Tamura M, Hayaishi O, Watanabe Y (1991) A possible mechanism of mitochondrial dysfunction during cerebral ischemia: inhibition of mitochondrial respiration activity by arachidonic acid. Arch Biochem Biophys 289: 33–38

58. Terada LS, Willingham IR, Rosandich ME, Leff JA, Kindt GW, Repine JE (1991) Generation of superoxide anion by brain endothelial cell xanthine oxidase. J Cell Physiol 148: 191–196

59. Tombaugh GC, Sapolsky RM (1990) Mild acidosis protects hippocampal neurons from injury induced by oxygen and glucose deprivation. Brain Res 506: 343–345

60. Tombaugh GC, Sapolsky RM (1993) Evolving concepts about the role of acidosis in ischemic neuropathology. J Neurochem 61: 793–803

61. Vlessis AA, Widener LL, Bartos D (1990) Effect of peroxide, sodium, and calcium on brain mitochondrial respiration in vitro: potential role in cerebral ischemia and reperfusion. J Neurochem 54: 1412–1418

62. Wagner KR, Kleinholz M, Myers RE (1989) Delayed neurologic deterioration following anoxia: brain mitochondrial and metabolic correlates. J Neurochem 52: 1407–1417

63. Wagner KR, Kleinholz M, Myers RE (1990) Delayed decreases in specific brain mitochondrial electron transfer complex activities and cytochrome concentrations following anoxia/ischemia. J Neurol Sci 100: 142–151

64. Warner DS, Smith M-L, Siesjö BK (1987) Ischemia in normo- and hyperglycemic rats: effects on brain water and electrolytes. Stroke 18: 464–471

65. Zaidan E, Sims NR (1993) Selective reductions in the activity of the pyruvate dehydrogenase complex in mitochondria isolated from brain subregions following forebrain ischemia in rats. J Cereb Blood Flow Metab 13: 98–104

66. Zaidan E, Sims NR (1994) The calcium content of mitochondria from brain subregions following short-term forebrain ischemia and recirculation. J Neurochem 63: 1812–1819

Correspondence: Bo K. Siesjö, M.D., Laboratory for Experimental Brain Research, Experimental Research Center, Lund University Hospital, S-221 85 Lund, Sweden.

Acta Neurochir (1996) [Suppl] 66: 15–20
© Springer-Verlag 1996

Is Calcium Accumulation Post-Injury an Indicator of Cell Damage?

R. Linde[1], **H. Laursen**[2], and **A. J. Hansen**[1]

[1] Department of Neuropharmacology, Novo Nordisk A/S, and [2] Laboratory of Neuropathology, University Hospital, Copenhagen, Denmark

Summary

It is generally agreed that excessive intracellular calcium accumulation is the main culprit for nerve cell damage following brain injury. Many autoradiographic studies of the post-injury brain have demonstrated an accumulation of $^{45}Ca^{2+}$ in regions exhibiting neuronal damage. We have recently observed, after cortical contusion trauma [10], that there was a discrepancy between the extent of cell damage and the extent of $^{45}Ca^{2+}$ in autoradiograms; rather the distribution of $^{45}Ca^{2+}$ followed that of serum proteins. In addition $^{45}Ca^{2+}$ was also observed in white matter, which had no signs of damage. We tested the hypothesis that $^{45}Ca^{2+}$ accumulation was coupled to the presence of protein by directly injecting albumin into the brain cortex. There was a highly significant correlation between the content of $^{45}Ca^{2+}$ and of albumin as measured by ELISA. A similar pattern was found after a cortical freeze-lesion in the contralateral hemisphere. However, in the ipsilateral hemisphere where cell damage was observed, the relation broke down and calcium accumulated in excess. We conclude that calcium accumulation in the brain is not only the result of cell damage but also of the presence of calcium-binding proteins, e.g. albumin.

Keywords: Albumin; $^{45}Ca^{2+}$; freeze-lesion; secondary brain damage.

Introduction

It is generally believed that calcium has a pivotal role in ischemic nerve cell death. According to this "calcium hypothesis" excessive loading of the neuron by Ca^{2+} leads to cell death by activation of various enzymes including proteases, endonucleases, etc. (e.g. Siesjö and Bengtsson [18]. The concept is built on a number of observations made in non-neuronal tissues. In 1965 McLean *et al.* [7], observed calcium accumulation by livers which have been exposed to toxins, and Schanne *et al.* [14] reported that the ability of toxins to damage liver cells in primary cultures was dependent on the presence of extracellular calcium. The hypothesis was implemented for the central nervous system by Siesjö [17]. Thereafter a plethora of papers have emerged

demonstrating that calcium did accumulate when the nerve cells histologically exhibited signs of injury. $^{45}Ca^{2+}$ has been extensively studied in the post-injury brain (e.g. Dienel [1]), in which the isotope is injected into the blood and circulates for 5 h before autoradiography of the brain is carried out. We have used a similar technique in the post-trauma brain and have confirmed that calcium accumulates when cell damage was present [10]. However, close inspection of the sections revealed that the area of calcium accumulation by far exceeded the "volume" of damaged cells. When the brain sections were stained for serum proteins the concordance with the calcium accumulation was striking. We, therefore, suggested that the accumulation of calcium in the post-traumatic brain was not only the result of uptake into damaged cells but also the result of protein accumulation.

The purpose of the present experiments was to test this hypothesis. We used a quantitative method for determining albumin in brain tissue in order to disclose a possible relationship between accumulation of calcium and accumulation of proteins (i.e. albumin).

Materials and Methods

Infusion of Albumin into the Brain

Male Sprague-Dawley rats (n=3) weighing 350–400 g were anaesthetized with pentobarbital sodium (50 mg/kg, i.p.). The head of the rat was fixed in a stereotactic frame and a midline scalp incision was made to expose the skull over both hemispheres. A 1 mm burr hole was drilled above each hemisphere at three positions 3 mm laterally to the midline: bregma +3.0, –2.1, –6.9 mm. The dura was left intact. In one hemisphere solutions with a different albumin content were infused at the three sites whereas the contralateral hemisphere received only saline. The infusion system consisted of a pointed glass pipette with a tip diameter of approx. 50 µm connected

16

R. Linde *et al.*

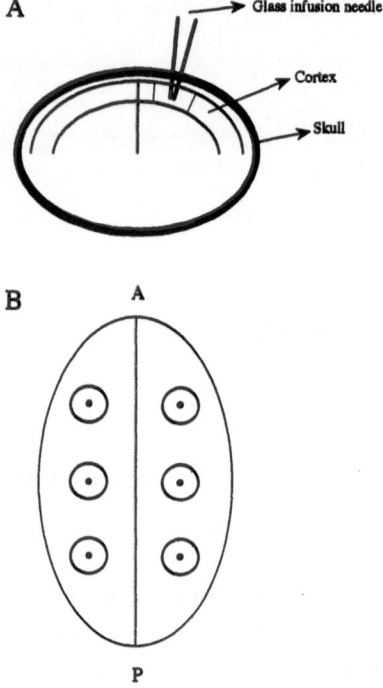

Fig. 1. Schematic drawing of the technique used for albumin infusion into brain cortex of the rat. A pointed glass pipette with a diameter of approx. 50 μm is connected to a syringe filled with either saline or saline with albumin (A). The lower part shows the six infusion sites, the three on the right side received albumin and the three on the left received saline. The larger circles around the infusion site correspond to the tissue samples. *A* is anterior and *P* posterior (B)

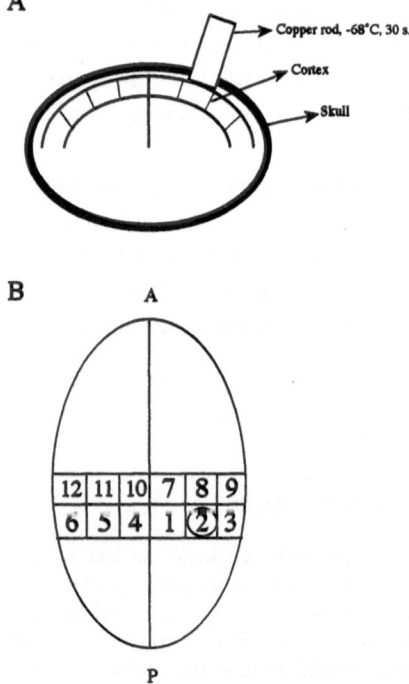

Fig. 2. Schematic drawings showing the principles of the freeze-lesion to the right hemisphere of the rat (A) and how the 12 brain samples are obtained by two coronal sections. The circle at tissue sample 2 is the frozen area. *A* is anterior and *P* posterior (B)

with a polyethylene tubing to an insulin syringe (0.1 ml) placed in an infusion pump. The syringe contained either sterile physiological saline or saline with albumin added (20 mg/ml rat serum albumin, essentially fatty acid – and globulin free (A 6414, Sigma St. Louis, U.S.A.)). The glass pipette was attached to a micromanipulator and placed 2 mm below the surface of the brain for each burr hole before infusion. The rate of infusion was always 3 μl/min but volumes varied from 4–30 μl. Following the infusion procedure the skin incisions were sutured and the rats were kept anaesthetized. Two hours later a polyethylene catheter was introduced into the jugular vein in order to infuse 100 μCi ^{45}Ca^{2+} (Dupont de Nemours, Belgium) in 1 ml physiological saline over 2 min. The catheter was subsequently removed. Five hours later the rats were reanaesthetized, blood samples were taken by heart puncture, the rats were decapitated and the brains removed. The brain cortex surrounding each infusion site was isolated by using a 4-mm cork bore, and the tissue samples were placed in preweighed vials (Fig. 1).

Cortical Freeze-Lesion

Male Sprague-Dawley rats (n=3) weighing 250–300 g were anaesthetized with pentobarbital sodium (50 mg/kg, i.p.). The head was fixed in a stereotactic frame, the skin above the skull incised, and the galea removed. A 6 mm trephination was made over the right parietal cortex using a dental drill. Special care was taken not to injure the dura. The freeze-lesion was made by a cylindrical copper probe (5 mm OD) precooled to –60 °C by a dry-ice/acetone mixture [6,22]. The probe was lowered onto the exposed intact dura using a micromanipulator and left there for 30 seconds (Fig. 2). Thereafter the bone fragment was replaced, the skin wound closed by sutures and the animals let awake in cages with free access to water and food. Twenty hours later the rats were reanaesthetized using pentobarbital sodium (50 mg/kg, i.p.) and a polyethylene catheter was placed in the jugular vein. Approximately 100 μCi ^{45}Ca^{2+} (Dupont de Nemours, Belgium) in 1 ml saline was infused over 2 min and allowed to circulate for 5 hours. After this period the rats were reanaesthetized, blood was sampled by heart puncture in heparinised syringe and centrifuged to obtain plasma. The rats were subsequently decapitated and the brains carefully removed. Two coronal sections (3–4 mm in width) were cut manually; one frontal to the lesion site and one including the lesion. The cortex of each section was divided into three ipsilateral and three contralateral samples, resulting in 12 samples per brain (Fig. 2).

Measurement of Albumin Content and Calcium Uptake

The samples were placed in preweighed vials and the wet weight determined. Triton X-405, (5 %) (Sigma, St. Louis, U.S.A.) was added to a PBS-Tween solution consisting of 0.05 % Tween 20 (Sigma St. Louis, U.S.A.) in phosphate buffered saline. This solution diluted the brain samples 200-fold (wt/vol) before sonication using a Branson Sonifier 250. The homogenates were left for 2 hours at room temperature before centrifugation. The content of albumin in the supernatant was determined by an enzyme-linked immunosorbent assay (ELISA) and the ^{45}Ca^{2+} content by scintillation counting. The ELISA method was adopted from Menzies *et al.* [8] with some modifications. In principle the method uses an antibody to albumin that is absorbed to plastic and that binds the albumin of the test samples to the well. The bound albumin is then detected using the same antibody conjugated to an enzyme, the reaction product of which can be visualized. We used a 96-well polystyrene plate (NuncImmunoPlate PolySorp F96, Nunc A/S, Denmark) and each well was filled with 100 μl 4 % sheep IgG to rat albumin (Cappel cat.# 55729, NC, U.S.A.) in carbonate/bicarbonate buffer (0.1 *M*;

pH=9.6), covered using sealing tape (Nunc A/S, Denmark) before incubation overnight at 4 °C. The wells were washed 3 times in PBS-Tween; each time for 3 minutes. This washing procedure was repeated after each step in the assay. Wells were then filled with 5 % nonfat dry milk powder in PBS-Tween and incubated for 2 hours at room temperature. After washing, duplicate wells were filled with 100 µl of a solution of purified rat albumin (fraction V, A 6272, Sigma, St.Louis, U.S.A.) in PBS-Tween at 80.0, 40.0, 20.0, 10.0, 5.0, 2.5, 1.25, 0.625 and 0.312 ng/ml concentrations to obtain the standard curve. Similarly, 100 µl of different dilutions of the supernatant of brain homogenate were filled in duplicate wells. In separate wells all reagents except the antigen was added to establish the background value. Plates were then covered by sealing tape (Nunc A/S, Denmark) and stored at 4 °C overnight. After washing, the wells were filled with 100 µl of sheep IgG to rat albumin conjugated to peroxidase (Cappel cat.# 55776, NC, U.S.A.) and incubated for two hours at room temperature. After washing, 100 µl substrate solution containing 0.67 mg orto-phenylenediamine (OPD)(S 2000 DAKO A/S, Denmark) per ml 0.1 M citric acid/phosphate buffer pH=5.0, and 5 µl 30 % hydrogen peroxide (mixed immediately before use) was added. Each well was exposed exactly 10 min in the dark for color development before the reaction was arrested by adding 150 µl 1 M sulphuric acid. The amount of yellow/brown reaction product was determined by measuring the absorbance at 490 nm using an ELISA-reader (ImmunoReader NJ-2000, Teknunc A/S, Denmark). Standard curves were drawn and values from the different dilutions were plotted on the standard curve, and averaged to calculate tissue albumin content.

The $^{45}Ca^{2+}$ contents of the brain homogenate supernatant and plasma are determined by scintillation counting.

Statistical analysis was carried out by linear regression analysis.

Histology

Three additional rats with a freeze-lesion were used for histological evaluation. Twenty hours after injury the rats were reanaesthetized using pentobarbital sodium (50 mg/kg, i.p.) and perfusion fixed with 4 % buffered formalin. The brains were postfixed in the same fixative for 24 hours and subsequently dehydrated and embedded in paraffin. Four µm sections were stained with hematoxylin-eosin or immunohistochemically with peroxidase/antiperoxidase technic for glial fibrillary acidic protein (GFAP) or for serum proteins.

Results

All the rats survived the operational procedure. In both experimental groups the calcium uptake was measured by the uptake of $^{45}Ca^{2+}$ in brain. In order to account for differences in levels of $^{45}Ca^{2+}$ in the blood the values of the brain samples were normalized by using the level of $^{45}Ca^{2+}$ in the plasma obtained immediately before decapitation.

Albumin Infusion

Direct infusion of albumin in the brain cortex increased the content of albumin. The increase was accompanied by an increase of $^{45}Ca^{2+}$ content. As shown in Fig. 3 there was a highly significant correlation (r=0.8610; n=21; p<0.0001) between the amounts of albumin and calcium in the brain cortex. In this set of data are included the results obtained from the contralateral site which received saline (unfilled circles). The average content was 0.4 mg/g wet wt which is in agreement with control values reported by Szymas et al. [21] and Menzies et al. [8]

Freeze-Lesion

The cortical freeze-lesion caused dramatic changes in the contents of albumin and $^{45}Ca^{2+}$ in brain cortex when measured after one day. The increases of both

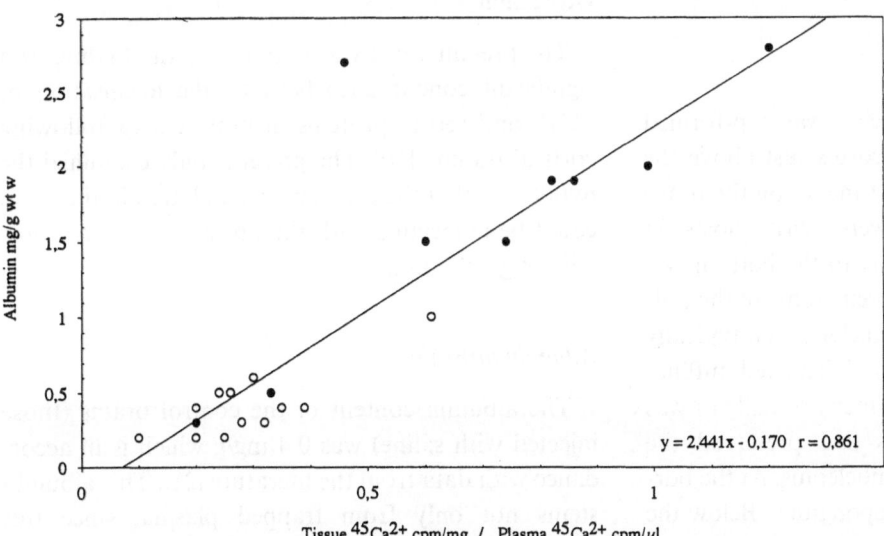

y = 2,441x - 0,170   r = 0,861

Tissue $^{45}Ca^{2+}$ cpm/mg / Plasma $^{45}Ca^{2+}$ cpm/µl

Fig. 3. Scattergram showing the relationship between the brain cortical albumin content and the brain cortical $^{45}Ca^{2+}$ content measured 7 h after cortical infusion of rat serum albumin (see methods). The brain $^{45}Ca^{2+}$ content is expressed as tissue counts divided by the end plasma counts. Unfilled circles represent the saline infused tissue samples, filled circles represent albumin infused tissue samples. The correlation between the amount of albumin and calcium is highly significant (r=0.8610; n=21; p<0.0001)

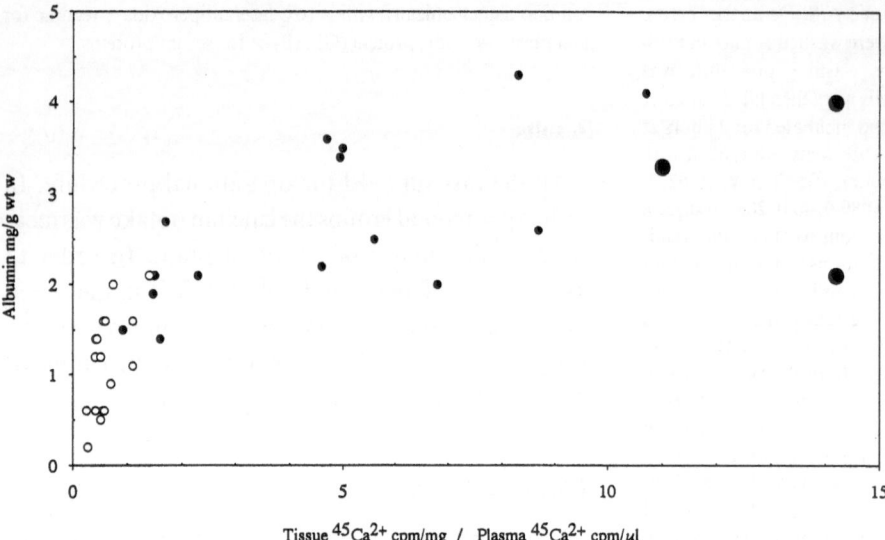

Fig. 4. Scattergram showing the relationship between the brain cortical albumin content and the brain cortical ⁴⁵Ca²⁺ content after a cortical freeze-lesion in rats performed 20 h earlier. The brain $^{45}Ca^{2+}$ content is expressed as tissue counts divided by the end plasma counts. Unfilled circles represent the cortex samples from the contralateral hemisphere, filled circles the ipsilateral hemisphere in which the big filled circles represent the samples from the freezed region

albumin and $^{45}Ca^{2+}$ were far more extensive in the traumatized hemisphere than in the other, but the levels of the contralateral hemisphere were markedly increased as well. Fig. 4 shows the brain albumin content in relation to brain $^{45}Ca^{2+}$. The lower region of the curve, which represents the contralateral tissue samples and the tissue samples most distant from the infarct of the traumatized hemisphere, shows that the brain albumin content increased linearly with the content of $^{45}Ca^{2+}$ in the brain. When the albumin content in the brain exceeds approx. 3 mg/ml the curve levels off and reaches a plateau where only the content of $^{45}Ca^{2+}$ increases.

Histology

In all three rats the freeze-lesion was cup-formed reaching to the bottom of the cortex just above the corpus callosum. In the center of the lesion the tissue was coagulated and the neurons were seen as ghosts. At the edge and reaching all the way to the bottom two types of neuronal damage were seen. Some of the cells were eosinophilic with pyknotic nuclei, as it is typically seen after ischemia. Other cells had distorted outlines with retracted cytoplasm that stained normally or was slightly basophilic. These cells had preserved the nuclear structure with a distinct nucleolus. At the border of the lesion the tissue was spongious. Below the lesion there was an extensive edema in the corpus callosum. The edema was reaching across the midline into the corpus callosum of the contralateral site. With GFAP stain two of the three brains showed a very

slight increase in GFAP reactivity at the border of the lesion, whereas the third showed no changes. In the serum protein stains there was an extensive reaction in and around the freeze-lesion and all the way through the corpus callosum across the midline, but apparently no reaction in the cortex. This is in contrast to the results obtained with ELISA, but may be explained by a lower sensitivity of the immunostain. In the contralateral cortex no neuronal damage, astrocytic reactions or serum protein extravasation studied by immunohistochemical analysis was encountered.

Discussion

The present work was initiated by the finding of a significant concordance between the localisation of $^{45}Ca^{2+}$ and serum proteins in brain cortex following cortical trauma [10]. The present study examined the hypothesis that the calcium accumulation in the brain could be associated with the presence of serum proteins, e.g. albumin.

Albumin in Brain

The albumin content of the control brains (those injected with saline) was 0.4 mg/g which is in accordance with data from the literature [21]. This albumin stems not only from trapped plasma, since this amounts to 0.07 mg/ml (with a plasma albumin concentration of 14 mg/ml and a plasma space in brain of 0.005 mg/ml) but also from albumin bound to the endothelial cell surface [15] as well as albumin in the brain parenchyma. In cerebrospinal fluid the concentration is normally at 0.07 mg/ml, hence, the amount in brain tissue is probably minimal.

Calcium in the Brain

The method for measuring calcium accumulation in the brain was adopted from Dienel [1]. The method assumes that $^{45}Ca^{2+}$ given i.v. reaches a steady-state in different brain compartments after a 5 hours exposure time. The total amount of calcium in brain tissue is 1mmol/kg·s. Since the interstitial concentration is approx. 1 mmol/l and the interstitial space occupies approx. 20 % (e.g. Hansen [4], the amount of calcium in the interstitial space is about one fifth of the total calcium, i.e. in steady-state one fifth of the radioactive calcium is located in the interstitial space.

It was noted by studying autoradiograms of brains exposed to various circulation times of $^{45}Ca^{2+}$ that calcium preferentially entered the brain via the choroid plexus and was distributed via the ventricles [1]. When the technique was applied in animals at various times following an episode of severe global ischemia the autoradiograms revealed calcium accumulation in regions displaying neuronal damage, thereby supporting the theory of calcium involvement in ischemic cell damage.

Relationship Between Albumin and Calcium in Brain

We increased the albumin content locally by direct injection into the brain and found an increased content of $^{45}Ca^{2+}$. There was a highly significant linear correlation between $^{45}Ca^{2+}$ accumulation and albumin content. After all this should not be surprising since albumin binds calcium. In the plasma, calcium is present in three compartments, a free fraction, a complexed fraction and one protein bound fraction of which albumin accounts for 80 %. About one third of the total plasma calcium is bound to albumin. The question is whether the accumulated $^{45}Ca^{2+}$ is associated with albumin or whether other mechanisms take place. We have attempted to answer this question by comparing the ratio of $^{45}Ca^{2+}$ and albumin in plasma with their ratio in brain. In Table 1 are shown the values of the brain ratio divided by the plasma ratio for the two experimental groups. If all the accumulated calcium in brain was bound to albumin the ratio should have a value of around 5 based on the following reasoning. We assume that the albumin injected into the brain is located in the interstitial space whereas the

Table 1. *Fraction of $^{45}Ca^{2+}$ Bound to Brain Albumin*

	Sample		Rat I	Rat II	Rat III
Freeze-lesion					
	Infarct	(2)	50.6	54.1	33.9
	Ipsilateral	(1)	37.2	17.9	13.2
	–	(3)	8.9	9.1	21.1
	–	(7)	27.5	8.7	14.7
	–	(8)	20.3	27.1	33.7
	–	(9)	10.3	4.9	7.8
	Contralateral	(4)	9.7	7.7	3.8
	–	(5)	5.3	3.4	3.0
	–	(6)	5.0	7.7	3.2
	–	(10)	9.4	–	7.9
	–	(11)	13.1	5.6	3.4
	–	(12)	14.7	11.1	4.2
			Rat IV	Rat V	Rat VI
Infusion	albumin		6.9	7.8	3.1
	albumin		6.9	9.7	8.3
	albumin		9.9	16.0	9.0
	saline		9.0[a]	15.7	9.5
	saline		7.1[a]	13.8	11.0
	saline		9.7[a]	12.5	17.5

The table shows the values of the ratio of $^{45}Ca^{2+}$/albumin in plasma divided by the ratio of $^{45}Ca^{2+}$/albumin in brain after freeze-lesion in three rats (upper part). The lower part shows similar values from the brain cortex of three rats subjected to intracortical infusion of albumin in saline or saline (see methods). Sample # refers to the numbers in Fig. 2. [a]Indicates that only a craniotomy was performed

$^{45}Ca^{2+}$ during the 5 hours distributes according to the prevalent calcium amounts meaning that only one fifth is confined to the interstitial space. The values representing the injected animals and the contralateral sites of the freeze-lesioned rats are in line with this supposition albeit the scatter is large. The ipsilateral site of the freeze-lesion, however, and especially the site of the freezing exhibits a significantly higher ratio indicating that the accumulated calcium also is localized at other places than bound to albumin, e.g. necrotic tissue.

This means that when albumin enters the "normal" brain there is a proportional increase of calcium due to albumin binding. When the tissue is damaged as well there is an additional increase of calcium without relation to albumin.

There are other studies which suggest a relationship between albumin and calcium accumulation. When the method of Dienel [1] was used in rats after middle cerebral artery occlusion (MCAO) there was a progressive accumulation of $^{45}Ca^{2+}$ which seemed to peak 3 days after ictus in the affected cortex [9,16]. This is hard to explain, if accumulation of $^{45}Ca^{2+}$ should indicate cell damage since the cortical cells die within 6 hours after MCAO. However, the fact that the albumin content of the affected cortex displays a similar time course reaching a peak after 3 days [8] could explain the finding.

Possible Neurotoxicity of Albumin

There is evidence that albumin itself is neurotoxic. In cell cultures of cerebellar granule cells albumin not only causes cell death but also potentiates the toxicity of glutamate [2]. Furthermore albumin injected directly into the caudate nucleus of rats leads to cell damage when examined 5 days later [5]. Other studies have demonstrated a relationship between opening of the BBB and the presence of tissue damage. Thus, when the BBB was opened either by carotid infusions of hyperosmolar solutions [12,13] or by experimental hypertension [19] and after experimental seizures [20] there was structural damage in the regions with opening of BBB. It was stated that the pathological mechanism of the irreversible injury was related to the extravasation of plasma constituents but the mechanism remained unclear. An important role of albumin seemed more obvious for inducing damage of the thalamus which is seen in the aftermath of middle cerebral artery occlusion in rats. The lesion develops when albumin and probably other plasma constituents have migrated from the brain infarct in cortex via white matter to the thalamus [11]

An explanation for the purported toxicity of albumin is not at hand. Eimerl *et al.* [2] have suggested that albumin keeps the NMDA gated channel open, thereby facilitating entry of calcium. This could be mediated by complexing free Zn-ions which in the normal state decreases the open state probability of the NMDA channel [3]. In other words albumin could enhance the toxic effect of glutamate. If this statement is valid for the brain *in vivo*, it would put special emphasis on conditions in which the BBB breaks down. The presence of albumin alone may be rather innocuous but in conditions of an increased glutamate release the possibility for neuronal cell damage is greatly increased. Thus, the integrity of the blood-brain barrier may be quite important for the development of cell damage after brain injury.

References

1. Dienel GA (1984) Regional accumulation of calcium in postischemic rat brain. J Neurochem 43: 913–925
2. Eimerl S, Schramm M (1991) Acute glutamate toxicity and its potentiation by serum albumin are determined by the Ca^{2+} concentration. Neurosci Lett 130: 125–127
3. Eimerl S, Schramm M (1993) Potentiation of ^{45}Ca uptake and acute toxicity mediated by the N-methyl-D-aspartate receptor: the effect of metal binding agents and transition metal ions. J Neurochem 61: 518–525
4. Hansen AJ (1985) Effect of anoxia on ion distribution in the brain. Physiol Rev 65: 101–148
5. Hassel B, Iversen EG, Fonnum F (1994) Neurotoxicity of albumin in vivo. Neurosci Lett 167: 29–32
6. Klatzo I, Piraux A, Laskowski EJ (1958) The relationship between edema, blood-brain barrier and tissue elements in local brain injury. J Neuropath Exp Neurol 17: 548–564
7. McLean AEM, Mclean E. Judah JD (1965) Cellular necrosis in the liver induced and modified by drugs. Int Rev Exp Pathol 4: 127–157
8. Menzies SA, Betz AL, Hoff JT (1993) Contributions of ions and albumin to the formation and resolution of ischemic brain edema. J Neurosurg 78: 257–266
9. Nagasawa H, Araki T, Kogure K (1992) Autoradiographic analysis of second messenger and neurotransmitter receptor bindings in the strionigral system of the postischemic rat brain. J Neurosci Res 33: 485–492
10. Nilsson P, Laursen H, Hillered L, Hansen AJ (1996) Calcium movements in traumatic brain injury: the role of glutamate receptor operated ion channels. J Cereb Blood Flow Metab 16: in press
11. Nordborg C, Sokrab TEO, Johansson BB (1991) The relationship between plasma protein extravasation and remote tissue changes after experimental brain infarction. Acta Neuropathol 82: 118–126
12. Salahuddin TS, Johansson BB, Kalimo H, Olsson Y (1988) Structural changes in the rat brain after carotid infusions of hyperosmolar solutions: a light microscopic and immunohistochemical study. Neuropathol Appl Neurobiol 14: 467–482
13. Salahuddin TS, Kalimo H, Johansson BB, Olsson Y (1988) Observations on exudation of fibronectin, fibrinogen and albumin in the brain after carotid infusions of hyperosmolar solutions. Acta Neuropathol 76: 1–10
14. Schanne FAX, Kane AB, Young EA, Farber JL (1979) Calcium dependence of toxic cell death: a final common pathway. Science 206: 700–702
15. Schnitzer JE, Oh P (1994) Albondin-mediated capillary permeability to albumin. J Biol Chem 269: 6072–6082
16. Shirotani T, Shima K, Iwata M, Kita H, Chigasaki H (1994) Calcium accumulation following middle cerebral artery occlusion in stroke-prone spontaneously hypertensive rats. J Cereb Blood Flow Metab 14: 831–836
17. Siesjö BK (1981) Cell damage in the brain: a speculative synthesis. J Cereb Blood Flow Metab 1: 155–185
18. Siesjö BK, Bengtsson F (1989) Calcium fluxes, calcium antagonists, and calcium-related pathology in brain ischemia, hypoglycemia, and spreading depression: A unifying hypothesis. J Cereb Blood Flow Metab 9: 127–140
19. Sokrab T-EO, Johansson BB, Kalimo H, Olsson Y (1988) A transient hypertensive opening of the blood-brain barrier can lead to brain damage. Acta Neuropathol 75: 557–565
20. Sokrab T-EO, Kalimo H, Johansson BB (1990) Parenchymal changes related to plasma protein extravasation in experimental seizures. Epilepsia 31: 1–8
21. Szymas J, Hossmann K-A (1990) Determination of endogenous serum proteins in normal and oedematous brain tissue of cat by rocket and crossed immunoelectrophoresis . Acta Neurochir (Wien) 105: 169–177
22. Unterberg A, Schmidt W, Dautermann C, Baethmann A (1990) The effect of various steroid treatment regimens on cold-induced brain swelling. Acta Neurochir (Wien) [Suppl] 51: 104–106

Correspondence: A.J. Hansen, M.D., Neuropharmacology, Health Care Discovery, Novo Nordisk A/S, Novo Nordisk Park, 2760 Måløv, Denmark.

Acta Neurochir (1996) [Suppl] 66: 21–26

Apoptosis in Focal Cerebral Ischemia

M. Chopp[1,2] and **Y. Li**[1]

[1]Department of Neurology, Henry Ford Health Science Center, Detroit, MI and [2]Department of Physics, Oakland University, Rochester, MI, U.S.A.

Summary

An ischemic insult to the brain evokes cell damage which may progress to cell death. We invariably associate cell death with necrosis. Necrosis exhibits well defined morphological characteristics, and the biochemical and biophysical processes associated with necrosis have been identified. However, another form of cell death exists, apoptosis. Apoptosis plays an important role in the early development of tissues. Cells undergoing apoptosis exhibit very different morphological characteristics and temporal profiles of change from cells undergoing necrosis. Apoptosis has been identified with the internucleosomal fragmentation of DNA. More importantly, apoptosis has been associated with a process of programmed cell death, in which a genetic program is activated which results in the death of the cell. In this presentation, we will review our data on the morphological, biochemical and molecular evidence of apoptosis in the rodent (rat, mouse) brain after middle cerebral artery occlusion. Emphasis will be placed on describing the temporal profile and the anatomical distribution of cells undergoing apoptosis as functions of duration of MCA occlusion and reperfusion after MCA occlusion. The possible contribution of selective genes in promoting and/or inhibiting apoptosis will also be discussed.

Keywords: Apoptosis; anatomical distribution; temporal profile; focal cerebral ischemia.

History and Background

The term, apoptosis, coined in 1972 by Kerr, derives from Greek and refers to falling off, as leaves fall from a tree [18]. Morphological characteristics of apoptosis include compaction of chromatin against the nuclear membrane, cytoplasm shrinkage with preservation of organelles, and nuclear and cytoplasmic budding to form membrane-bound fragments, referred to as apoptotic bodies. In contrast, necrosis involves initial swelling of the cell followed by a breakdown of cell membranes and disintegration of nuclear structure and cytoplasmic organelles. The cell then shrinks, condenses and disintegrates. Although Kerr made specific reference to a new form of cell death [17], the morphological changes associated with apoptosis predate those of Kerr's by nearly 100 years. In 1886, a botanist, Berthold, asserted that protein denaturation was responsible for certain forms of cell death [2]. In 1885, Walter Flemming showed the half moons of pyknotic chromatin, typical of apoptosis [9]. In the early 1950s, Glücksmann described a form of cell death specific to the embryo [11], and the concept of cell suicide arose with the discovery of lysosomes. An excellent review on the history of apoptosis can be found in Majno and Joris [29].

The actual death of the cell may precede necrosis or apoptosis. That is, from morphological characteristics we cannot identify when the cell has crossed the boundary to irreversible injury or death. Death is a functional term, which has not been clearly defined. After irreversible injury a cell may still function and be metabolically active, yet may be mortally injured without exhibiting necrotic or apoptotic morphology.

A biochemical marker which has become the hallmark of apoptosis is that of DNA fragmentation. Wyllie in 1980 found a ladder pattern on gel electrophoresis of thymocytic apoptosis [40]. The DNA was segmented at 180 bp intervals reflecting the activity of endonuclease cleavage of DNA at internucleosomal sites. The identity of the endonuclease is still a mystery. Although gel electrophoresis indicating DNA degradation is routinely identified with apoptosis [1], DNA fragmentation may not always correlate with apoptotic morphology and is a conventional marker and not equivalent to morphologically identified apoptosis [38]. It remains to be resolved whether DNA digestion is a specific marker for apoptosis in vivo [15].

A molecular biological-histochemical method has recently been developed which permits morphological detection of DNA fragmentation [10]. This method combines two complementary indices of apoptosis, the DNA fragmentation and the morphological signature of apoptotic bodies, to identify apoptotic cells. 3'-OH DNA ends are generated by DNA fragmentation and are typically localized to morphologically identifiable apoptotic bodies and nuclei. Residues of digoxigenin-nucleotide are catalytically added to the DNA by terminal deoxynucleotidyl transferase, an enzyme which catalyzes a template independent addition of deoxyribonucleotide triphosphate to the 3'-OH ends of double- or single-stranded DNA. (This method is referred to as TUNEL staining.) We have employed this method in our studies to localize the DNA fragmentation in ischemic brain [21–24].

Apoptosis has been closely associated with programmed cell death (PCD), a term coined by Lockshin and Williams [27]. PCD is a functional term, used to describe cell death that is a normal part of the life of a multicellular organism. Apoptosis, on the other hand, is a descriptive term, to describe a type of cell death exhibiting a distinct set of morphological features. An excellent review of this topic can be found in Schwartz and Osborne [36]. It is important to clarify the concept of programmed cell death. There are two ways, not mutually exclusive, of defining this term. PCD can refer to the timing of onset of events that lead to cell death. This type of cell death has been investigated in great detail in developmental biology ([31], review). It occurs during metamorphosis [13]. Extensive PCD also occurs during the early development of the nervous system and is regulated by trophic factors ([31], review).

Another interpretation of PCD is that of an orchestrated sequence of gene expression resulting in the eventual death of the cell. A major catalyst for conceptual framework in this area derives from investigations performed on the nematode *Caenorhabditis (C.) elegans* [8,13,42]. This roundworm contains 1,090 somatic cells of which 131 die at specific stages of development and undergo a genetically controlled death. Three distinct programmed phases of killing have been identified in this system; 1) killing of the cell; 2) phagocytosis of the corpse; 3) digestion of the engulfed constituents. Mutations in specific genes can block each of these events. The genes are referred to as ced (cell death abnormal) genes. Mutations in ced-3 and ced-4 block PCD and the 131 cells survive. Mutations in ced-2, ced-5 and ced-10 block phagocytosis.

There may be killer genes, analogous to ced-3 and ced-4, which when blocked promote survival. It is interesting to note, that the ced-3 gene has a similar sequence to a cysteine protease, interleukin 1β-convertase, and cysteine proteases may promote cell death ([39], review). In contrast to the 'killer gene', there are protector genes in the *C. elegans* system. ced-9 protects cells from PCD. Transfecting cells with ced-9 inhibit the death of the 131 cells destined to die. The ced-9 has an analogue in mammals, the bcl-2 gene ([32], review). Cell lines transfected with bcl-2 are resistant to killing agents, e.g. withdrawal of growth factors, heat shock, glucocorticoids and γ-rays, to name a few [33]. bcl-2 can replace ced-9 in roundworms to protect against PCD. A caveat to these important data indicating a genetic component to the process of cell death, is that not all apoptosis requires de novo gene expression or the synthesis of new proteins. DNA degradation and loss of transcription cannot completely explain apoptosis since lens cells and erythrocytes undergo apoptosis, and enucleated cytoplasts can undergo apoptosis and bcl-2 administration protects these cells from apoptosis ([36], review). Likewise, apoptosis can occur in the absence of new gene expression, i.e. cytotoxic T lymphocytes (CTL) induced apoptosis. DNA fragmentation can also occur within 20 minutes, a time too short for transcription [30,34]. Conversely, not all PCD occurs by apoptosis. The tobacco hawk moth loses segmental muscles by a PCD, which requires de novo gene expression, but does not display membrane budding or chromatin margination or DNA fragmentation [13].

It is important to realize that genes may be upregulated in dying cells, yet they may not be essential for cell death. Other genes are essential for specific forms of cell death, and there are distinct signal transduction pathways that mediate PCD with different stimuli. For example, T-cells in p53 knockout mice display apoptosis when exposed to glucocorticoids or T-cell receptor (TCR) stimulus; however, they fail to die when exposed to γ-radiation [6,28].

Apoptosis and PCD in Focal Cerebral Ischemia

Apoptosis has traditionally been associated with physiological events such as: embryonic development, CTL killing and tumor regression. Tissue subjected to injury and insult also undergoes apoptosis, as is apparent from the pioneering studies of ischemic liver of Kerr and Wyllie [17,18].

In our studies, we employed a model of middle cerebral artery occlusion induced by insertion of a 4-O

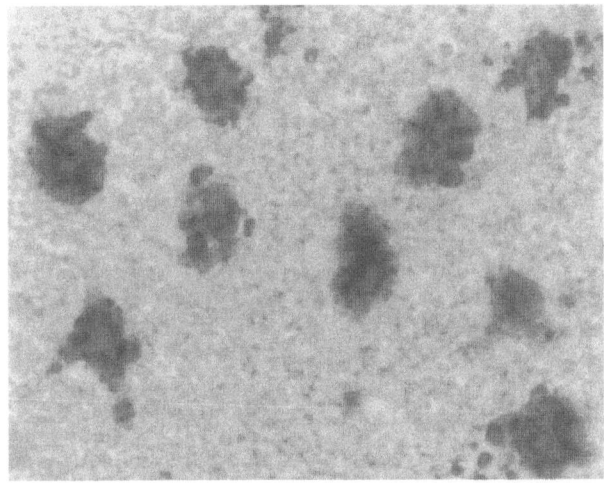

Fig. 1. Apoptotic cells within the ischemic striatum of a rat subjected to 2 h of MCA occlusion and 24 h of reperfusion. Apoptotic cells are identified by means of TUNEL stain (ApopTag kit, Oncor, Inc., Gaithersburg, MD)

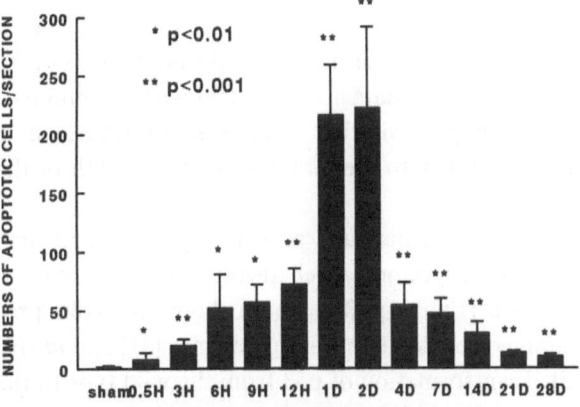

Fig. 2. Apoptosis after MCA occlusion. Apoptotic cells in coronal sections at the level of the anterior commissure in rats subjected to 2 h of middle cerebral artery (MCA) occlusion at 0.5 h to 28 days of reperfusion. (Reprinted with permission from [21])

nylon monofilament into the internal carotid artery to occlude the MCA in both the mouse and the rat [43]. At various times after transient MCA occlusion, the brains were prepared by perfusion fixation with heparinized saline and 10 % buffered formalin phosphate. In both the mouse [23] and the rat [21,22,24] apoptotic cells were readily observed within the ischemic area. Figure 1 illustrates the apoptotic cells present within the ischemic striatum of a rat subjected to 2 hours of MCA occlusion and 24 hours of reperfusion. Using double staining techniques, neuron specific enolase for identification of neurons, glial fibrillary acidic protein (GFAP) for astrocytes, and Factor VIII as a marker for endothelial cells, in combination with the TUNEL stain, we determined that 90–95 % of cells exhibiting apoptotic morphology were neurons, followed by as-

trocytes (~5–10 %) and endothelial cells (≤1 %). This suggests that the sensitivity of cells exhibiting apoptotic morphology reflects the selective vulnerability of cells undergoing necrosis after focal cerebral ischemia [19]. Complementary data to the in situ detection of apoptosis in ischemic brain has recently been provided in ultrastructural studies in samples from ischemic tissue [22].

We measured the distribution of apoptotic cells after various durations of reperfusion after 2 hours of MCA occlusion in the rat [21]. Figure 2 shows the numbers of apoptotic cells obtained from ischemic rat brain as a function of time of reperfusion. The numbers of apoptotic cells peaked at 24 and 48 hours after ischemia and then subsequently declined. However, the numbers of apoptotic cells remained significantly elevated compared to control animals, even 4 weeks after MCA occlusion. Thus, the presence of apoptosis weeks after the onset of ischemia indicates that cell death is an ongoing dynamic process, and that interventions even weeks after the onset of the ischemic event may salvage compromised cells. Since the half-life of apoptotic cells may be short compared to necrotic cells, simply counting the relative numbers of apoptotic cells compared to necrotic cells may underestimate the contribution of apoptosis to ischemic cell damage. We estimate that the numbers of apoptotic cells to be approximately 4 % at 24 and 48 hours after onset of ischemia. Thus, it is likely that apoptosis may be responsible for far more than 4 % of the cells that die as a result of the ischemic insult.

The distribution of apoptotic cells within the ischemic tissue as a function of reperfusion time is shown in Fig. 3. The vast majority of the apoptotic cells are located along the inner boundary of the ischemic core of the lesion [21,24]. The reason for this preferential spatial distribution within the ischemic lesion is not known. However, the localization of apoptotic cells at the inner boundary of the ischemic lesion suggests that the apoptotic process contributes to the expansion of the ischemic lesion. Apoptosis has also been associated with free radical damage [3,14,16], and it is possible that cells at the boundary of the ischemic lesion adjacent to viable tissue with uncompromised blood flow may be particularly susceptible to free radical damage evoked by reperfusion injury. The selective distribution of apoptotic cells within the ischemic zone may imply that there are distinct pathophysiological mechanisms or initiating events responsible for necrosis and apoptosis.

The numbers of apoptotic cells also increase as a function of duration of ischemia [24]. Figure 4 illus-

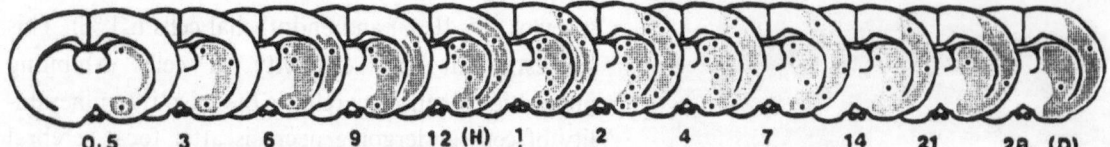

Fig. 3. Apoptosis and neuroual necrosis-rat MCA. Illustration of the temporal profile and distribution of apoptotic cells (dots, ApopTag kit) and necrotic neurons (hatched, hematoxylin and eosin). Although scattered apoptotic cells increase throughout the territory of the MCA, groups of apoptotic cells are localized primarily to the inner boundary zone of the infarction, both in the stratum and in the cortex at 24–48 h of reperfusion. (Reprinted with permission from [21])

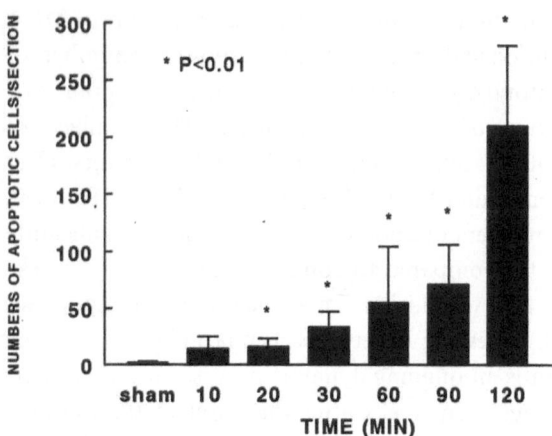

Fig. 4. Apoptosis after MCA occlusion. Progressive increase in numbers of apoptotic cells with increasing times from 10–120 min in the territory of MCA. (Reprinted with permission from [24])

trates the numbers of apoptotic cells present within the ischemic lesion as a function of duration of ischemia, from 10 minutes to 2 hours. All animals were sacrificed at 48 hours of reperfusion. The numbers of apoptotic cells clearly increase as a function of duration of ischemia. These data suggest that even brief durations of ischemia, i.e. 10 minutes, that evoke few necrotic cells, induce apoptosis. Thus, mild ischemic injury can cause apoptosis. The degree of apoptotic injury reflects the intensity of the ischemic event, with longer durations of ischemia causing increasing numbers of apoptotic cells. These data also suggest that apoptosis and necrosis are complementary and parallel events. It is however not known, whether necrosis and apoptosis are mutually exclusive events. Whether cells can enter into an apoptotic pathway and subsequently become necrotic, or the converse is not known.

Until now our focus has been on the morphological evidence of apoptosis in ischemic brain. Apoptosis has been closely associated with the process of programmed cell death, and the upregulation of death genes and death proteins [36]. Although there are counter examples from the literature in which apoptosis is present without upregulation of new genes

and where new gene expression is present without the morphological evidence of apoptosis, the preponderance of studies have closely linked apoptosis with either gene expression or protein translation.

Linnik et al. have demonstrated that administration of cycloheximide to rats subjected to middle cerebral artery occlusion significantly reduces ischemic cell damage [25]. The inhibition of protein synthesis by administration of cycloheximide disrupts the processes of programmed cell death and hence reduces apoptosis. They have also shown that the DNA fragmentation as measured by gel electrophoresis is most prominent in the tissue located at the boundaries of the insult [26]. These data are consistent with our findings, and the hypothesis that apoptosis contributes in a substantial way to the maturation and growth of the ischemic lesion.

Certain proteins such as wild-type (wt) p53 may promote programmed cell death and contribute to apoptosis [12]. wt p53 has been shown to evoke programmed cell death in tumor cell lines [41] and in vivo [35]. Transformation of p53 from the wild type to the mutant form induces cancer in many cell and organ systems by inhibiting programmed cell death. We have detected p53 in ischemic rat brain subjected to MCA occlusion [5,20]. Figure 5 shows an immunohistochemical section of ischemic rat brain stained for wild type p53. The arrows show p53 positive cells also exhibiting apoptotic bodies and thus apoptosis. p53 positive cells are present without morphological evidence of apoptosis (arrow heads). Thus, it is possible that p53 may be a necessary but not sufficient condition for apoptosis in ischemic brain. Recent studies of MCA occlusion in p53 knockout mice demonstrate that ischemic cell damage is reduced in p53 transgenic knockout mice [7], indicating that p53 may play a role in the promotion of ischemic cell death after MCA occlusion. Many other genes and proteins have been associated with the promotion of cell death [36], and the p53 data in ischemic brain are not meant to exclude the possibility that other genes contribute to this process.

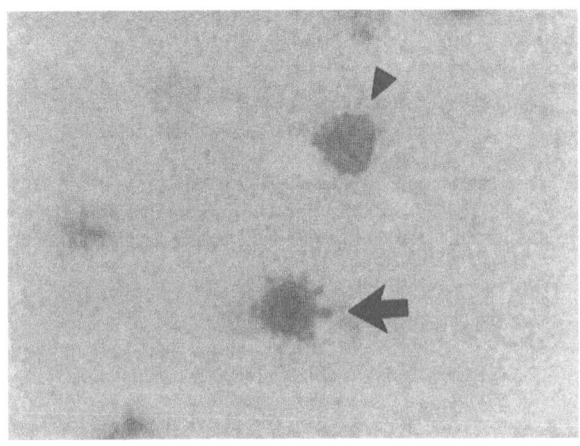

Fig. 5. Immunohistochemical section of ischemic rat brain stained for wild-type p53. The arrow shows p53 positive cells also exhibiting apoptotic bodies and thus apoptosis. p53 positive cells are present without morphological evidence of apoptosis (arrow head)

In addition to apoptosis and programmed cell death contributing to the net ischemic cell death, it is likely that the flip side of programmed cell death, DNA repair may promote cell survival. p53 has been shown to play an important role in DNA repair. p53 is a transcription factor and turns on another gene, GADD45 (growth-arrest-and-DNA-damage-inducible) [37]. GADD45 complexes with proliferating cell nuclear antigen (PCNA), a necessary component of the machinery that copies DNA. Investigations of programmed cell death and apoptosis in cerebral ischemia should therefore be coupled to the possibility of DNA repair as a therapeutic intervention after stroke.

In summary, apoptosis is a well established morphological complement or alternative to necrosis. It occurs in many biological systems and is reponsible for embryonic development, and physiological events such as metamorphosis and immunological responses. The role played by apoptosis after injury to tissue is an area of vital interest. Investigations of cell damage after focal cerebral ischemia should include the likelihood that apoptosis contributes to ischemic cell damage after focal cerebral ischemia. Apoptosis has been associated with gene and molecular alterations. This opens up new avenues of therapy based on the possibility of intervention with processes regulating cell death as well as events contributing to cell repair.

References

1. Arends MJ, Morris RG, Wyllie AH (1990) Apoptosis. The role of the endonuclease. Am J Pathol 136: 593–608
2. Berthold G (1886) Studien über Protoplasmamechanik. Verlag Arthur Felix Leipzig
3. Buttke TM, Sandstrom PA (1994) Oxidative stress as a mediator of apoptosis. Immunol Today 15: 7–10
4. Cameron GR (1951) Pathology of the cell. Thomas, Springfield
5. Chopp M, Li Y, Zheng ZG, Freytag SO (1992) p53 expression in brain after middle cerebral artery occlusion in the rat. Biochem Biophys Res Comm 182: 1201–1207
6. Clark AR, Purdie CA, Harrison DJ, Morris RG, Bird CC, Hooper ML, Wyllie AH (1993) Thymocyte apoptosis induced by p53-dependent and independent pathways. Nature 362: 849–852
7. Crumrine RC, Thomas AL, Morgan PF (1994) Attenuation of p53 expression protects against focal ischemic damage in transgenic mice. J Cereb Blood Flow Metab 14: 887–891
8. Ellis HM, Horvitz HR (1986) Genetic control of programmed cell death in the nematode C. elegans. Cell 44: 817–829
9. Flemming W (1885) Über die Bildung von Richtungsfiguren in Säugethiereiern beim Úntergang Graafscher Follikel. Arch Anat Entw Gesch 221–244
10. Gavrieli Y, Sherman Y, Ben-Sasson SA (1992) Identification of programmed cell death in situ via specific labeling of nuclear DNA fragmentation. J Cell Biol 119: 493–501
11. Glücksmann A (1951) Cell deaths in normal vertebrate ontogeny. Biol Rev Camb Philos Soc 26: 59–86
12. Gottlieb E, Haffner R, von Rüden T, Wagner EF, Oren M (1994) Down-regulation of wild-type p53 activity interferes with apoptosis of IL-3 dependent hematopoietic cells following IL-3 withdrawal. EMBO J 13: 1368–1374
13. Hedgecock EM, Salston JE, Thomson JN (1983) Mutations affecting programmed cell deaths in the nematode Caenorhabditis elegans. Science 220: 1277–1279
14. Hockenbery DM, Nunez G, Milliman C, Scheriber RD, Korsmeyer SJ (1990) Bcl-2 is an inner mitochondrial membrane protein that blocks programmed cell death. Nature 348: 334–336
15. Hockenbery D (1995) Defining apoptosis. Am J Pathol 146: 16–19
16. Kane DJ, Sarafian TA, Anton R, Hahn H, Gralla EB, Valentine JS, Ord T, Bredesen DE (1993) Bcl-2 inhibition of neural death: decreased generation of reactive oxygen species. Science 262: 1274–1277
17. Kerr JFR (1971) Shrinkage necrosis: a distinct mode of cellular death. J Pathol 105: 13–20
18. Kerr JFR, Wyllie AH, Currie AR (1972) Apoptosis: a basic biological phenomenon with wide-ranging implications in tissue kinetics. Br J Cancer 26: 239–257
19. Li Y, Chopp M, Garcia JH, Yoshida Y, Zhang ZG, Levine SR (1992) Distribution of the 72-kd heat shock protein as a function of transient focal cerebral ischemia in rats. Stroke 23: 1292–1298
20. Li Y, Chopp M, Zhang ZG, Niewenhuis L, Gautam S (1994) p53-immunoreactive protein and p53 mRNA expression after transient middle cerebral artery occlusion in the rat. Stroke 25: 849–856
21. Li Y, Chopp M, Jiang N, Yao F, Zaloga C (1995) Temporal profile of in situ DNA fragmentation after transient middle cerebral artery occlusion in the rat. J Cereb Blood Flow Metab 15: 389–397
22. Li Y, Sharov VG, Jiang N, Yao F, Zaloga C, Sabbah HN, Chopp M (1995) Ultrastructural and light microscopic evidence of apoptosis after middle cerebral artery occlusion in rat. Am J Pathol 146: 1045–1051
23. Li Y, Chopp M, Jiang N, Zaloga C (1995) In situ detection of DNA fragmentation after focal cerebral ischemia in mice. Molec Br Res 28: 164–168

24. Li Y, Chopp M, Jiang N, Zhang ZG, Zaloga C (1995) Induction of DNA fragmentation after 10–120 minutes of focal cerebral ischemia in rat. Stroke 26: 1252–1258

25. Linnik MD, Zobrist RH, Marsa D, Hatfield MD (1993) Evidence supporting a role for programmed cell death in focal cerebral ischemia in rats. Stroke 24: 2002–2009

26. Linnick MD (1995) Personal communication

27. Lockshin RA, Williams CM (1965) Programmed cell death. I. Cytology of degeneration in the intersegmental muscles of the Pernyl silkmoth. J Insect Physiol 11: 123–133

28. Lowe SW, Schmitt EM, Smith SW, Osborne BA, Jacks T (1993) p53 is required for radiation-induced apoptosis in mouse thymocytes. Nature 362: 847–849

29. Majno G, Joris I (1995) Apoptosis, oncosis, and necrosis. An overview of cell death. Am J Pathol 146: 3–15

30. Martz E, Howell DM (1989) CTL: virus control cells first and cytolytic cells second ? DNA fragmentation, apoptosis and the prelytic halt hypothesis. Immunol Today 10: 79–86

31. Oppenheim RW (1991) Cell death during development of the nervous system. Ann Rev Neurosci 14: 453–501

32. Raff MC, Barres BA, Burne JF, Coles HS, Ishizaki Y, Jacobson MD (1993) Programmed cell death and the control of cell survival: lessons from the nervous system. Cell 262: 695–700

33. Reed JC (1994) Bcl-2 and the regulation of programmed cell death. J Cell Biol 124: 1–6

34. Russell JH, Dobos CB (1980) Mechanisms of immune lysis II. CTL-induced nuclear disintegration of the target begins within minutes of cell contact. J Immunol 125: 1256–1261

35. Schimke RT, Mihich E (1994) Fifth annual Pezcoller symposium: apoptosis. Cancer Res 54: 302–305

36. Schwartz LM, Osborne BA (1993) Programmed cell death, apoptosis and killer genes. Immunol Today 14: 582–590

37. Smith ML, Chen I-T, Zhan Q, Bae I, Chen C-Y, Gilmer TM, Kastan MB, O'Connor PM, Fornace AJ Jr (1994) Interaction of the p53-regulated protein Gadd-45 with proliferating cell nuclear antigen. Science 266: 1376–1380

38. Tomei LD, Shapiro JP, Cope FO (1993) Apoptosis in C3H/10T/1/2 mouse embryonic cells: evidence for internucleosomal DNA modification in the absence of double-strand cleavage. Proc Natl Acad Sci USA 90: 853–857

39. Vaux DL, Haecker G, Strasser A (1994) An evolutionary perspective on apoptosis. Cell 76: 777–779

40. Wyllie AH (1980) Glucocorticoid-induced thymocyte apoptosis is associated with endogenous endonuclease activation. Nature 284: 555–556

41. Yonish-Rouach E, Resnitzky D, Lotem J, Sachs L, Kimchi A, Oren M (1991) Wild-type p53 induces apoptosis of myeloid leukaemic cells that is inhibited by interleukin-6. Nature 353: 345–347

42. Yuan J, Horvitz HR (1990) The Caenorhabditis elegans genes ced-3 and ced-4 act autonomously to cause programmed cell death. Dev Biol 138: 33–41

43. Zea Longa E, Weinstein PR, Carlson S, Cummins R (1989) Reversible middle cerebral artery occlusion without craniectomy in rats. Stroke 20: 84–91

Correspondence: Michael Chopp, Ph.D., Department of Neurology, Henry Ford Hospital, 2799 West Grand Blvd., Detroit, MI 48202, U.S.A.

Acta Neurochir (1996) [Suppl] 66: 27–31

Significance of the Inflammatory Response in Brain Ischemia

J. M. Hallenbeck

National Institute of Neurological Disorders and Stroke, National Institutes of Health, Bethesda, MD, U.S.A.

Summary

Leukocytes appear to have a central role in the inflammatory response that develops during acute brain ischemia. This brief review adduces evidence that leukocytes accumulate in focal zones of acute brain ischemia at a sufficiently early stage to participate in the process of progressive ischemic brain damage and that partial inhibition of that accumulation, by various measures, can attenuate ischemic brain injury. Mechanisms of leukocyte adhesion are discussed in detail and an inference is put forward that leukocytes are an important factor in progressive ischemic injury, but almost certainly act in concert with a number of other similarly important factors. On this basis, leukocyte inhibition may have demonstrable benefit in acute stroke, but ultimately be found to only partially spare potentially salvageable tissue in the ischemic zone.

Keywords: Leukocytes; inflammation; adhesion receptors; ischemia; stroke.

The principles governing blood flow through a zone of acutely ischemic brain have, until recently, been viewed by workers in the stroke field as being closely approximated by Ohm's Law applied to the circulation. In that view, the ratio of pressure to resistance determined the rate of delivery of oxygen and glucose to the injured tissue and these factors were sufficient to describe the contribution of the circulation to the eventual amount of brain damage. In recent years, the situation has become far more complex with the recognition that blood is capable of undergoing a multifactorial interaction with ischemically-damaged brain tissue and that, in the course of this interaction, multiple inflammatory mediators are generated that can contribute to the progression of injury [18].

Leukocytes are a central feature of inflammation and the delayed accumulation of these cells in reactive zones around infarcts has been well-established in histopathologic specimens from human autopsies [16], but these cells have been classically considered to func-

tion as a cleanup crew that prepares the wound for healing. In the last few years, many laboratories have reported that leukocyte accumulation in ischemically-injured brain is an early event that coincides with the period of damage progression and could, therefore, participate in the process. In 1986, a time course of leukocyte accumulation in acute brain ischemia was determined [17]. Autologous granulocytes labelled with ^{111}In were infused into a dog model of cerebral ischemia induced by incremental air embolism and monitored by the cortical somatosensory evoked potential. Accumulation of leukocytes was measured by gamma scintigraphy and visualized relative to cerebral blood flow by double-label autoradiography. Following one hour of ischemia by this technique, leukocytes were observed to accumulate from 1 to 4 hours into the recirculation period particularly in zones of impaired reflow. A potential limitation of this study has been that air emboli directly damage endothelium [21] and this could alter the dynamics of leukocyte accumulation as compared with vascular occlusion models in which vessels are indirectly activated as a consequence of the vascular stasis. In a well-established model of middle cerebral artery occlusion (MCAO) in the awake baboon, polymorphonuclear leukocyte plugging of up to 40% of capillaries was observed after 3 hours of ischemia and 1 hour of reperfusion [13]. The authors concluded that the polymorphonuclear neutrophils (PMNs) contributed to vessel damage and progressive impairment of microvascular perfusion in that model. In an intravascular suture model of permanent MCAO in the wistar rat, PMNs were detected by light and electron microscopy in capillaries and venules within 30 min and monocytes appeared after 4–6 hrs in these same vessels [15]. This represents the earliest detection of PMN and monocyte influx in focal brain ischemia.

Pioneering studies of endothelial cells and leukocytes in culture systems during the last 15 to 20 years have contributed to a detailed understanding of the biochemical and cellular mechanisms involved in leukocyte activation, adhesion and transendothelial migration. The attachment of circulating blood cells to endothelium involves a variety of complementary receptors and is orchestrated by a host of signalling molecules. The description that follows is based on several recent reviews [5–7]. Two major sets of complementary molecules control adhesive interactions between circulating cells and endothelium, selectins that recognize carbohydrate ligands and members of the immunoglobulin (Ig) superfamily that recognize the integrins.

The selectin family was first recognized in 1989 after a series of reports described three cell surface glycoproteins on ECs, platelets, and leukocytes that had related cDNA sequences. According to a standard nomenclature which designates each family member according to the cell type on which it was originally identified, these surface glycoproteins have been designated E-selectin (endothelium), P-selectin (platelets), and L-selectin (lymphocytes).

The selectin family members all share structural homologies. Each is characterized by an N-terminal lectin-like domain, epidermal growth factor (EGF) repeats, and a series of complement binding protein modules. Expression of L-selectin is constitutive; that of E-selectin is inducible with a peak at 4–6 hrs and a return to baseline by 24–48 hrs. Preformed P-selectin is a transmembrane protein located in alpha granules and dense granules of platelets and in Weibel-Palade bodies of ECs. It is translocated to the cell surface in minutes after effective stimulation. P-selectin can also be synthesized *de novo* and expressed over the course of hours in response to cytokine stimulation. Selectins bind via their lectin and EGF domains to one or more types of carbohydrate ligands including structures related to sialylated Lewis x (sLe^x) and sialylated Lewis a (sLe^a).

The other major set of bimolecular interactions that mediate adhesion of circulating blood cells to endothelium involves members of the Ig superfamily reacting with their complementary ligands, members of the integrin family. Members of the Ig superfamily are structurally similar in that they all have a variable number of extracellular Ig-like domains, a transmembrane domain and a short cytoplasmic sequence. There are four principal molecular species in this group, intercellular adhesion molecule-1 (ICAM-1), intercellular adhesion molecule-2 (ICAM-2), vascular adhesion molecule-1 (VCAM-1), and CD31. ICAM-1 is constitutively expressed at low levels by endothelium from most of the vascular segments assayed and it is upregulated by cytokines with maximal expression in 16–24 hrs and remains elevated for at least 72 hrs in the presence of continued stimulation. ICAM-2 is constitutively expressed on nonactivated endothelium at levels 10–15 times higher than that of ICAM-1 and cytokine exposure does not upregulate its expression. This receptor could be important for rapidly occurring inflammatory events. VCAM-1 is either minimally or not expressed on nonactivated endothelium and it is induced by cytokines with peak expression by 6–12 hrs. Its expression is also rapidly induced on aortic endothelium by an atherogenic diet [24]. One other recently identified receptor in the Ig superfamily is CD31 (PECAM-1, endoCAM) which tends to distribute near endothelial cell borders and may participate in EC-EC interactions that limit vascular permeability by virtue of CD31–CD31 homophilic adhesion. In addition, homophilic interaction between CD31 on leukocytes and CD31 on ECs may be important in transendothelial migration. CD31 can also undergo heterophilic adhesion with an unknown counter-receptor.

The term "integrin" as applied to a group of receptors was initially intended to emphasize that the receptors served to integrate signals from the extracellular environment with the intracellular cytoskeleton [23]. It turns out that this is a general property of circulating blood cell adhesion to ECs in that each interaction tends to involve intracellular signalling for participating cells in addition to the binding [26]. Integrins are a family of heterodimeric molecules in which each member is composed of an alpha subunit covalently linked to a beta subunit. The β_1 and β_2 subfamilies are particularly important for leukocyte adhesion. Very late after activation antigen-4 (VLA-4) from the β_1 subfamily reacts with VCAM-1 on cytokine-activated ECs. The β_2 subfamily includes three receptors that are active at the blood-endothelial interface and are composed of a CD11a, CD11b, or a CD11c alpha subunit that is covalently linked to a common CD18 (β_2) beta subunit. Regulation of integrin activity involves several mechanisms. Integrin-mediated cell-cell interaction requires the binding of divalent cations such as Ca^{2+} or Mg^{2+} to the cation-binding domains on the α-subunit of integrin molecules. Upregulation of surface expression of integrin molecules is controlled in at least two ways.

One involves *de novo* synthesis and surface expression. The other involves rapid (minutes) mobilization of integrin molecules from peroxidase-negative granules to the surface upon stimulation. Modulation of the affinity of integrin molecules for their respective ligands by phosphorylation of the β-subunit is another mechanism for controlling the adhesive process.

Despite the apparent welter of receptors and regulatory systems as studied *in vitro*, interactions between circulating cells and endothelium in the body seem to be smoothly orchestrated. L-selectin acts as a guide for unstimulated leukocytes and mediates the rolling of leukocytes along unstimulated endothelium. As a leukocyte encounters a segment of activated endothelium, there is a transient increase in the functional activity of L-selectin and the rolling slows or stops. This exposes the leukocyte to a local gradient of chemotactic factors for a sufficient time to become programmed for activation. As the cell is signalled by such chemoattractant molecules as PAF, interleukin-8 (IL-8), and granulocyte-macrophage colony-stimulating factor (GM-CSF), Mac-1 is mobilized from intracellular granules to the leukocyte surface and a process of L-selectin shedding begins. The subsequent upregulation of Mac-1 and LFA-1 and progressive shedding of L-selectin seem to be required for transendothelial migration. P-selectin may also be mobilized from intracellular granules to the EC surface in the very early stages of a vascular insult and may participate in the first wave of leukocyte recruitment as may E-selectin [1,22,28].

This knowledge has permitted development of antibodies directed at specific adhesion molecules such that partial inhibition of leukocyte accumulation in zones of ischemically-damaged brain can now be achieved. Investigators in many laboratories have now demonstrated that administration of these antibodies can reduce the volume of tissue damage in experimental models of brain ischemia. Intravenous administration of an anti-CD18 monoclonal antibody that inhibits PMN adherence to endothelium, IB4, 15 min before reperfusion in baboons subjected to 3 hr of MCAO and 1 hr of reperfusion, increased reflow in microvessels of all size classes as determined by India ink tracer perfusion [25]. The authors concluded that CD18-mediated PMN adherence to endothelium contributes to "no-reflow" in microvessels after focal cerebral ischemia. Another group reported that administration of anti-CD11b was associated with a 28 % reduction in lesion volume in rats subjected to 2 hr of intravascular suture-induced MCAO followed by 46 hr of reperfusion [10]. Lesion volume was determined by histological evalua-

tion of hematoxylin and eosin (H&E) stained sections. The antibody was serially administered at 1 hr of reperfusion and at 22 hr of reperfusion and compared with an identical dosage schedule of an isotype-matched control antibody. PMNs and monocytes within brain tissue were quantified by means of a myeloperoxidase assay and the numbers of intraparenchymal leukocytes were determined to be significantly reduced in the group receiving anti-CD11b antibody compared with the vehicle control group. The authors concluded that the anti-CD11b antibody administered 1 hr after the onset of reperfusion resulted in significant reduction of ischemic cell damage associated with a decrease of intraparenchymal leukocytes after transient (2-hr) focal cerebral ischemia in the rat. Another study from the same laboratory employed an identical model and furnished additional dose-response data that supported the same conclusion that the anti-CD11b antibody provided significant reduction of ischemic cell damage after transient focal cerebral ischemia in the rat and that, in addition, this reduction of damage was dose-dependent and associated with significant functional improvement [9].

Several groups have investigated the role of leukocytes in acute brain ischemia by severely reducing the number of circulating leukocytes prior to induction of the ischemia. In a rabbit model of thromboembolic stroke, neutrophils were reduced to about 16 % of the baseline value with anti-neutrophil anti-serum and another group of rabbits received anti-platelet anti-serum that reduced platelets to 23 % of the baseline value. There was also a control group that received non-immune serum. Four hr after embolization, the rabbits were killed and infarct size was determined by triphenyltetrazolium chloride (TTC). Neutropenic rabbits recovered their cerebral blood flow to a greater extent than animals in the control or thrombocytopenic groups and infarct size was significantly reduced in the neutropenic group only. The thrombocytopenic group did have a smaller increase in intracranial pressure than the control group, but the neutropenic group did not experience any increase in intracranial pressure during the experimental period [4]. The authors concluded that neutrophils may be important contributors to ischemia-induced brain injury, but that the role of platelets is probably more subtle. A potential criticism of this study is that they attempted to evaluate infarct size by means of TTC at a relatively early time point at which other investigators have questioned the reliability of the

TTC technique [3,19]. In rats subjected to 2 hr of transient MCAO by means of an intravascular suture that was subsequently withdrawn to provide 46 hr of reperfusion, a rabbit anti-rat PMN sera was infused immediately after withdrawal of the suture as reperfusion was initiated. Serial H&E-stained coronal brain sections were histologically evaluated and neutropenic animals had significantly smaller infarct volumes (19 %) than did the vehicle-treated controls (27 %) [8]. The authors concluded from the data that neutrophils may play an important role in mediating ischemic brain injury. Leukopenia induced by vinblastine with white blood cell counts of less than 1,500 per mm^3 has been reported to be associated with considerably smaller infarcts than normal levels of circulating leukocytes in a model of global forebrain ischemia. In this model, rats were subjected to bilateral carotid occlusion plus hypotension with blood pressure maintained at 50 mm Hg by blood withdrawal for a period of 60 min. Subsequently, the carotid arteries were unclamped and the blood was reinfused so that the animals were exposed to 75 min of reperfusion. Infarcts in leukopenic rats averaged 21 % versus a 70 % area of infarction in the control group. EEG was also preserved in all leukopenic animals when compared with controls both during ischemia and after reperfusion suggesting the possibility of better intra-ischemic perfusion in leukopenic animals [20]. The results were interpreted to indicate that white blood cells participate in the generation of cerebral damage in a model of global forebrain ischemia and reperfusion. Again, the use of TTC to delineate the cerebral infarcts in this study at such an early time point does raise a question about the reliability of the measurement technique since this procedure is regarded as more reliable after 24 hr than before 6 hr [3,19].

Leukocyte inhibition with a novel agent, doxycycline has also been shown to mitigate nerve cell damage in a rabbit model of spinal cord ischemia [12]. Doxycycline is a member of the teracycline family of antibiotics and it has been shown to inhibit in vitro human leukocyte superoxide synthesis, degranulation, and adherence to protein-coated surfaces [14].

Several groups have observed that leukocyte inhibition was ineffectual in reducing ischemic brain damage. In cats subjected to focal hemispheric ischemia by means of bilateral common carotid artery occlusion combined with unilateral MCAO for 90 min followed by 180 min of reperfusion, monoclonal antibody 60.3 directed against the CD18 adhesion moiety on PMNs and monocytes was administered intravenously between 40 min and 50 min of ischemia. The antibody had no effect on the recovery of cerebral blood flow, cortical somatosensory evoked potential, or infarct volume as quantified by TTC staining [27]. Anti-CD18 antibody administered to rabbits 30 min before inducing irreversible ischemia of the brain by means of intraarterial microspheres failed to reduce ischemic injury, but the same antibody produced a significant reduction in neurologic deficits in a reversible spinal cord ischemia model reported by the same group [11]. The authors concluded that treatment with leukocyte adhesion antibody reduces CNS ischemic injury in a reperfusion model but not in an irreversible occlusion model. The findings were felt to support the role of leukocytes as active participants in reperfusion injury. The "no-reflow" phenomenon occurs in the majority of gerbils after 30 min of bilateral occlusion of the common carotid arteries and 10 min of reperfusion as determined by India ink infusion. A group of gerbils rendered leukopenic by cyclophosphamide with 85 % reductions of their whole blood leukocyte count did not display a corresponding reduction in the incidence of "no-reflow" phenomenon when subjected to this protocol and compared with the control group that had normal levels of circulating leukocytes [2]. The authors concluded that the evidence from this study casts doubt on the hypothesis that leukocyte plugging plays a major role in the cerebral microcirculation's response to ischemia.

There are a number of leukocyte properties that are potentially conducive to ischemic and postischemic damage. These include the production of a variety of inflammatory mediators, generation of oxygen-centered free radicals, elaboration of granule-based toxins, impairment of blood rheology, acceleration of thrombosis and excitotoxin release. Despite the multifaceted injury potential that characterizes leukocytes, they probably function as a few strands in an intricately interwoven web of mediators and processes that, in aggregate, determine the progression of ischemic damage and their inhibition may be predicted to ameliorate, but not arrest, progressive brain damage in stroke.

Acknowledgement

The author wishes to thank Mrs. Mary Crawford for her excellent editorial assistance.

References

1. Abbassi O, Kishimoto TK, McIntire LV, Anderson DC, Smith CW (1993) E-selectin supports neutrophil rolling in vitro under conditions of flow. J Clin Invest 92: 2719–2730

2. Aspey BS, Jessimer C, Pereira S, Harrison MJG (1989) Do leukocytes have a role in the cerebral no-reflow phenomenon? J Neurol Neurosurg Psychiatry 52: 526–528

3. Bederson JB, Pitts LH, Germano SM, Nishimura MC, Davis RL, Bartkowski HM (1986) Evaluation of 2,3,5-triphenyltetrazolium chloride as a stain for detection and quantification of experimental cerebral infarction in rats. Stroke 17: 1304–1308

4. Bednar MM, Raymond S, McAuliffe T, Lodge PA, Gross CE (1991) The role of neutrophils and platelets in a rabbit model of thromboembolic stroke. Stroke 22: 44–50

5. Beekhuizen H, van Furth R (1993) Monocyte adherence to human vascular endothelium. J Leukoc Biol 54: 363–378

6. Bevilacqua MP (1993) Endothelial-leukocyte adhesion molecules. Annu Rev Immunol 11: 767–804

7. Bevilacqua MP, Nelson RM (1993) Selectins. J Clin Invest 91: 379–387

8. Chen H, Chopp M, Bodzin G (1992) Neutropenia reduces the volume of cerebral infarct after transient middle cerebral artery occlusion in the rat. Neurosci Res Comm 11: 93–99

9. Chen H, Chopp M, Zhang RL, Bodzin G, Chen Q, Rusche JR, Todd RF III (1994) Anti-CD11b monoclonal antibody reduces ischemic cell damage after transient focal cerebral ischemia in rat. Ann Neurol 35: 458–463

10. Chopp M, Zhang RL, Chen H, Li Y, Jiang N, Rusche JR (1994) Postischemic administration of an anti-Mac-1 antibody reduces ischemic cell damage after transient middle cerebral artery occlusion in rats. Stroke 25: 869–876

11. Clark WM, Madden KP, Rothlein R, Zivin JA (1991) Reduction of central nervous system ischemic injury in rabbits using leukocyte adhesion antibody treatment. Stroke 22: 877–883

12. Clark WM, Calcagno FA, Gabler WL, Smith JR, Coull BM (1994) Reduction of central nervous system reperfusion injury in rabbits using doxycycline treatment. Stroke 25: 1411–1416

13. del Zoppo GJ, Schmid-Schönbein GW, Mori E, Copeland BR, Chang C-M (1991) Polymorphonuclear leukocytes occlude capillaries following middle cerebral artery occlusion and reperfusion in baboons. Stroke 22: 1276–1283

14. Gabler W, Tsukuda N (1991) The influence of divalent cations and doxycycline on iodoacetamide-inhibitable leukocyte adherence. Res Commun Chem Pathol Pharmacol 74: 131–140

15. Garcia JH, Liu KF, Yoshida Y, Lian J, Chen S, del Zoppo GJ (1994) Influx of leukocytes and platelets in an evolving brain infarct (Wistar rat). Am J Pathol 144: 188–199

16. Graham DI, Brierley JB (1984) Vascular disorders of the central nervous system. In: Adams JH, Corsellis JAN, Duchen LW (eds) Greenfield's neuropathology. Wiley, New York, pp161–162

17. Hallenbeck JM, Dutka AJ, Tanishima T, Kochanek PM, Kumaroo KK, Thompson CB, Obrenovitch TP, Contreras TJ (1986) Polymorphonuclear leukocyte accumulation in brain regions with low blood flow during the early postischemic period. Stroke 17: 246–253

18. Hallenbeck JM (1995) Inflammatory reactions at the blood-endothelial interface in acute stroke. Adv Neurol

19. Hatfield RH, Mendelow AD, Perry RH, Alvarezs LM, Modha P (1991) Triphenyltetrazolium chloride (TTC) as a marker for ischaemic changes in rat brain following permanent middle cerebral artery occlusion. Neuropathol Appl Neurobiol 17: 61–67

20. Heinel LA, Rubin S, Rosenwasser RH, Vasthare US, Tuma RF (1994) Leukocyte involvement in cerebral infarct generation after ischemia and reperfusion. Brain Res Bull 34: 137–141

21. Johansson BB (1980) Cerebral air embolism and the blood-brain barrier in the rat. Acta Neurol Scand 62: 201–209

22. Kishimoto TK (1991) A Synamic model for neurotrophil localization to inflammatory sites. J Natl Inst Health Res 3: 75–77

23. Larson RS, Springer TA (1990) Structure and function of leukocyte integrins. Immunol Rev 114: 181–217

24. Li H, Cybulsky MI, Gimbrone MA Jr, Libby P (1993) An atherogenic diet rapidly induces VCAM-1, a cytokine-regulable mononuclear leukocyte adhesion molecule, in rabbit aortic endothelium. Arterioscler Thromb 13: 197–204

25. Mori E, del Zoppo GJ, Chambers JD, Copeland BR, Arfors KE (1992) Inhibition of polymorphonuclear leukocyte adherence suppresses no-reflow after focal cerebral ischemia in baboons. Stroke 23: 712–718

26. Singer SJ (1992) Intercellular communication and cell-cell adhesion. Science 255: 1671–1677

27. Takeshima R, Kirsch JR, Koehler RC, Gomoll AW, Traystman RJ (1992) Monoclonal leukocyte antibody does not decrease the injury of transient focal cerebral ischemia in cats. Stroke 23: 247–252

28. Von Andrian UH, Hansell P, Chambers JD, Berger EM, Torres Filho J, Butcher EC, Arfors, KE (1992) L-selectin function is required for β_2-integrin-mediated neutrophil adhesion at physiological shear rates in vivo. Am J Physiol (Heart Circ Physiol 32) 263: H1034–H1044

Correspondence: John M. Hallenbeck, M.D., National Institute of Neurological Disorders and Stroke, National Institutes of Health, Building 36, Room 4A03, 36 Convent Drive MSC 4128, Bethesda, Maryland 20892-4128, U.S.A.

Acta Neurochir (1996) [Suppl] 66: 32–39

Leukocytes, Macrophages and Secondary Brain Damage Following Cerebral Ischemia

M. Tomita and **Y. Fukuuchi**

Department of Neurology, School of Medicine, Keio University, Shinanomachi, Shinjuku-ku, Tokyo, Japan

Summary

The involvement of white blood cells in microvascular derangement as a cause of secondary brain damage following cerebral ischemia is reviewed. Relevant data from the literature are arranged in the chronological sequence of the microvascular derangement of the brain that occurs after cerebral arterial occlusion (as based on our own experimental observations). The inflammatory processes which appeared to be elicited by polymorphonuclear leukocytes (PMNL) in the ischemic region of the brain may begin with adhesion of PMNLs to endothelial cells, followed by blood-brain barrier disruption, transudation/exudation, edema, necrosis, and scar formation. Stimulated by cytokines released from damaged neurons and axons, two types of macrophages (ameboid and ramified) appear, increase in number in the ischemic lesion, and engulf the debris of dead neurons, degenerated axons. Further, macrophages may release cytokines which stimulate healing processes, such as astroglial proliferation and revascularization, and release neurotoxins which could gradually kill surviving neurons. Even under such circumstances, individual leukocytes/macrophages are well regulated by specific mediators/cytokines. An urgent task is thus to find ways of controlling these key mediators/cytokines to reduce the inflammatory process and the extent of neuronal death for attenuating the secondary brain damage, without altering their beneficial effects.

Keywords: Microglia; ischemic microvascular derangement; immune reaction; inflammatory changes.

Introduction

In a variety of brain disorders, including ischemic stroke, the extent of irreversible tissue damage appears to be dependent largely on the degree of terminal vascular insufficiency in the injured region. The development of such microcirculatory derangement might well represent a final common pathway, which could account for the augmentation of injury by many nonspecific factors. This article is a mini-review on the involvement of white blood cells in secondary brain damage. Up-to-date descriptions of relevant topics have appeared recently in two excellent overviews: one was provided by Kochanek and Hallenbeck [25] entitled "Polymorphonuclear leukocytes and monocytes/macrophages in the pathogenesis of cerebral ischemia and stroke", and the other by del Zoppo and Garcia [8] entitled "Polymorphonuclear leukocyte adhesion in cerebrovascular ischemia". The present review attempts to provide a broad outline of the white cell involvement in correlation with the various stages of ischemic microvascular derangement occurring in the brain. The overall framework for such microvascular derangement (Fig. 1) is based on observations [45,46, 52] obtained by our photoelectric method [51,53] and cranial window technique applied to the cat cerebral cortex which was subjected to ischemia by middle cerebral artery occlusion (MCAO). Details of the inflammatory microvascular derangement relate to the work of Kulka [26], who examined microvascular changes in the ear of rabbits which sustained cold injury. Data on the involvement of polymorphonuclear leukocytes (PMNL) and monocytes/macrophages in ischemic tissue of the brain were collected from the literature, and have been arranged arbitrarily according to each stage of the cerebral microvascular derangement.

Microvascular Derangement

Figure 1 illustrates the sequential stages of the ischemic microvascular derangement during evolution of an infarct following cerebral arterial occlusion.

Stage 0: The scheme at the top left of Fig. 1 shows a microvascular unit consisting of an artery-microvasculature-vein.

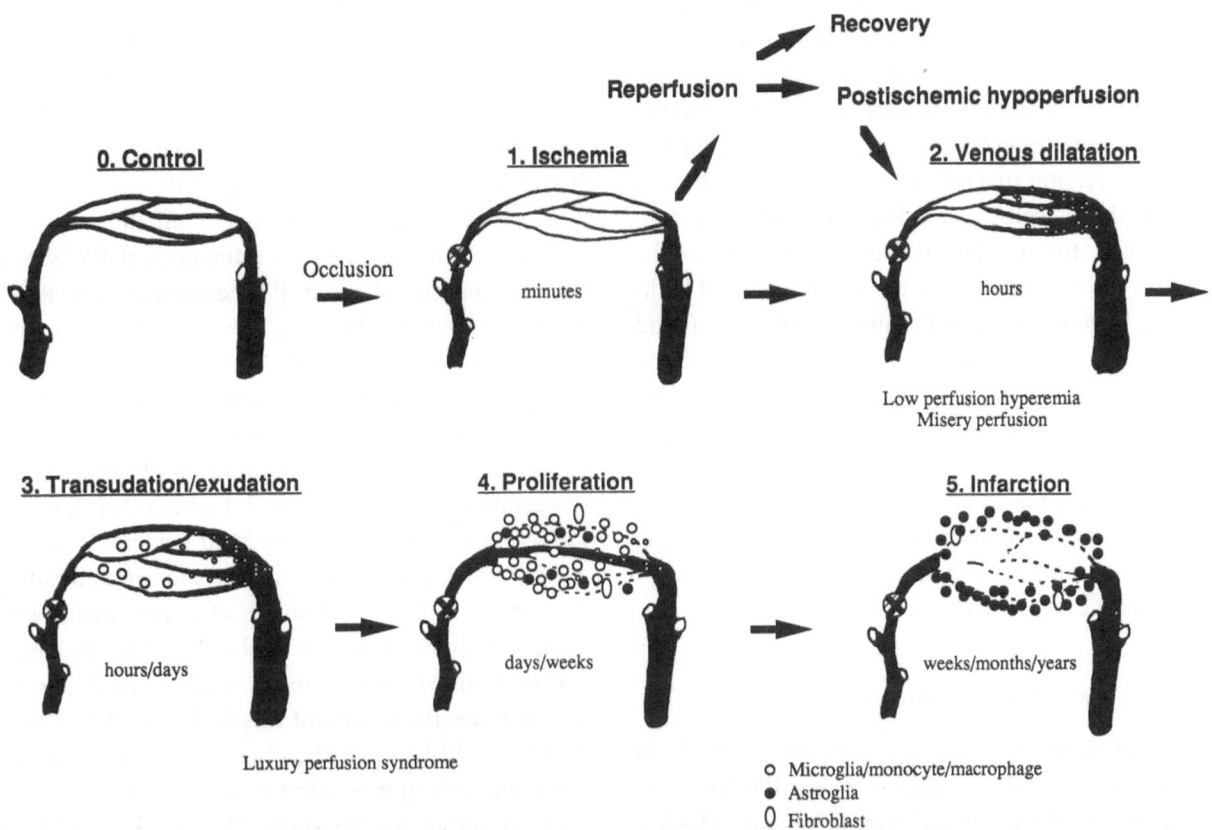

Fig. 1. Sequential stages of microvascular derangement of the brain. *Stage 0* a normal microvascular unit; *Stage 1* shrinkage of the microvasculature upon MCAO; *Stage 2* arterial constriction and venous dilatation; *Stage 3* glial swelling, BBB disruption, and edema; *Stage 4* red vein, and luxury perfusion; *Stage 5* infarct

Stage 1: Upon MCAO, the microvasculature shrinks. After stopping momentarily, the flow resumes sooner or later, and tends to recover in accordance with the development of collateral channels [45,46,52]. Leukocytes have been shown to participate in flow resistance through newly developing collateral channels even at an early stage due probably to their cellular shape [46,48,54]. Reperfusion by releasing the MCAO was found by Tanahashi [43] to produce various degrees of reactive hyperemia depending on the timing. He observed no change in CBV (cerebral blood volume) upon reopening within 1 min, and increases in CBV upon reopening after 1 min. Reperfusion produced multimodal individual time courses ranging from recovery after marked reactive hyperemia to severe postischemic hypoperfusion. No reflow was occasionally observed [43].

Stage 2: The CBV subsequently exceeds the preocclusive level (low perfusion hyperemia) due mainly to venous dilatation [24]. Direct observation reveals various changes in arterial diameter: simultaneous arterial dilation-constriction. This suggests a partial loss of vascular integrity, and a disruption of matching between flow and metabolism (e.g. misery perfusion syndrome).

Stage 3: By accumulation of surrounding extracellular fluid, the glial cells start to swell (cytotoxic edema) due to elevation of the osmotic potential in the cells (more than 200 mmHg) [50] induced by cell membrane damage. The neurons exhibit only slight swelling or even shrinkage due probably to partial rupture of the thin cell membrane [49]. The glial cells expand pushing aside surrounding structures and compressing the microvasculature. The blood-brain barrier (BBB) is disrupted with a delay of several hours after the occlusion. An increase in fluid permeability at the arterial side and absorption failure at the venous side of the microvasculature facilitate fluid retention in the tissue. Widening of the intercapillary distance, compression of vessels, focal leakage of plasma indicators, foggy, dark red areas of extravasated red cells, and marked aggregation of red cells are commonly observed. The experiments of postischemic hypoperfusion following reperfusion usually ran a similar course to that illustrating microvascular derangement after permanent occlusion.

Stage 4: Tissue ischemia becomes severer as indicated by a marked CBV decrease. Hemorrheological disturbances reach an irreversible level due to sludging of red blood cell clumps, platelet aggregation, plasma leakage, hemoconcentration, an increased blood viscosity, conversion to a rough procoagulant surface of the endothelial cells, and adhesion of platelets and leukocytes at the vessel wall. Spotty obliteration of the terminal vascular bed favors the development of high-pressure arteriolar-venular shunting (red vein and luxury perfusion syndrome). Carbon black injection reveals heterogeneous segmental stasis and low flow through the nutritional capillaries. These changes progress together until the central ischemic necrosis becomes well-demarcated.

Stage 5: All of the above described abnormal tissue changes subside resulting in a permanent defect with no-flow and cyst formation.

Mediators and White Cell Involvement

Stage 0: Little information is as yet available in clinical cases of stroke on whether or not PMNLs tend to be activated before the ischemic insult, and whether endothelial cells express adhesion molecules. A predisposition was demonstrated by Sirén *et al.* [39] with upregulation of ICAM-1 on SHR (spontaneous hypertensive rats) endothelial cells underlying the influence of hypertension on *in vivo* perivascular leukocyte accumulation in the parenchyma.

Stage 1: Again, there are still no sufficient data available, when and how PMNLs are activated, and when and how endothelial cells upregulate adhesion molecules during ischemia. Tanaka *et al.* [44] observed that after MCAO in cats, platelet thrombi were rapidly formed in cerebral microvessels irradiated with ultraviolet light, and that thrombus formation could be prevented by administration of a stable prostacyclin analogue (TRK-100). The rate of platelet aggregation has been shown by Rosenblum and El-Sabban [35] to be shear-rate dependent. Such early and continuous occurrence of platelet aggregation and release of mediators together with an increase in the TxA_2/PGI_2 ration may influence and/or trigger a cascade of events in the downstream microvasculature. A possible role for mediators released from platelets can be predicted from our data concerning the delay in recovery of CBV in platelet-depleted cats after MCAO [46,54]. Platelet aggregating factor (PAF), a chemical pro-inflammatory mediator, released from platelets, could be one of the candidates for such mediators to stimulate both

white cells and endothelial cells down-stream. Tanahashi *et al.* [42] found that prophylactic administration of PAF antagonist (TCV-309) suppressed postischemic hypoperfusion of the brain in cats. PAF as a mediator of specific pathological sequelae in stroke has been reviewed by Lindsberg *et al.* [27].

Stage 2: In general, the acute inflammatory response is switched on by the adherence of PMNLs to the vascular endothelium. PAF enhances adhesion of PMNLs to endothelium [3]. Tomita *et al.* have confirmed that activation of PMNLs by PAF is very rapid in an *in vitro* system [47], and that activated PMNLs adhering to rat and pig vascular endothelial cells are causing damage, such as laceration, shrinkage, and coagulation necrosis of the endothelium [58]. A speculative paradigm based on reported evidence could be as follows: Upon stimulation by such inflammatory mediators, as PAF, thrombin, etc., endothelial cells rapidly express adhesive molecules, e.g. P-selectin (GMP-140), which is stored in Weibel-Palade bodies, and initiates the release of interleukin (IL)-1, TNFα, and PAF. PMNLs in the circulating blood are also activated with upregulation of adhesive molecules on their cell surface as sialyl-Lewis X, Lewis X, and CD11/CD18. PMNLs are attracted by chemotactic agents, gather in the ischemic area, marginate in the microvessels (mostly venules), and roll on endothelial cells via P-selectin. The adhesion of PMNLs to the endothelium is mediated by a pair of conjugate adhesive molecules, CD11/18 and ICAM-1. The adherence can trigger an active oxygen radical generating system, such as xanthine oxidase in endothelial cells sufficient to injure these all elements [31]. Free radicals cause injury to cell membranes through a chain reaction peroxidation of lipids within the membrane structure and by enzyme inactivation [40]. Inflammatory mediators, such as IL-1, TNFα, and PAF are continuously released from the PMNLs, monocytes, as well as endothelial cells. Their targets are, again, PMNLs, monocytes, and endothelial cells. The positive feedback activation is increasing the endothelial damage in the tissue. TNFα has been reported to modulate endothelial cells to express hemostatic procoagulant properties [33].

Reperfusion injury involves multimodal events in which white cells are deemed to play a central role [21]. Upon reperfusion, toxic oxygen products, granular constituents of leukocytes, and phospholipase products become elevated in the tissue, since the reperfusing blood carries large amounts of white cells and oxygen. However, the reperfusion injury is based largely on the flow rate *per se* through the ischemically damaged tissue, no matter whether it is permanently occluded or reperfused. In either case, the inflammatory process undergoes various steps as follows.

Stage 3: A burst of superoxide radical production derived from both leukocytes and endothelial cells takes place. The free radicals damage adjacent endothelium and alter its function. Tissue damage is spreading in relation with the positive feedback mechanisms described above. PMNLs which were adhering to the endothelium slowly pass through the endothelial cell layer. Although Dorovini-Zis and Bowman [10] have noted that movement of PMNLs across the blood-brain barrier occurs by a transendothelial pathway without disruption of the barrier, the precise route remains to be defined. PMNLs emigrate further through the basement membrane with the aid of proteolytic enzymes. PAF, which has a short life-span in the tissue, is continuously released from the injured brain tissue together with eicosanoids, such as TxA_2, prostacyclin, and LTs (leukotrienes) due to enhanced lipid peroxidation. LTB_4, as one of the oxygenated derivatives of arachidonic acid, also displays pro-inflammatory effects, causing assembly, adhesion and chemotactic movement of PMNLs, and stimulating aggregation, enzyme release, and generation of superoxide in PMNLs [36]. Taken together, the releases of chemical mediators, arterial dilatation/constriction, venule dilatation, BBB damage, edema, cell membrane damage, and tissue damage form a cascade centering on PMNLs, leading to severe microcirculatory derangement and, thereby, secondary brain damage.

There is, in fact, an increasing body of evidence to indicate accumulation of PMNLs in ischemic lesions of the brain. In cats, Means and Anderson [28] have reported massive PMNL infiltration at 8 h after spinal cord compression, of macrophages at 8–24 h after the compression. del Zoppo *et al.* [9] made similar observations in baboons in a 3 h MCAO- and 1 h reperfusion model, and Hallenbeck *et al.* [22] in dogs with 60 min ischemia and 240 min postischemia by air embolism. Barone *et al.* [1] have noted 1) a dramatic accumulation of PMNLs in infarcted tissue of rats at 24 h after MCAO, 2) infiltration of PMNLs into the cerebral tissue based on analysis of hematoxylin-eosin stained histologic sections, in which PMNLs were identified at various stages of diapedesis into the brain parenchyma, and 3) an increase in myeloperoxidase activity in brain homogenates. Recently, Garcia *et al.* [13] reported that PMNLs could be detected in rat brain subjected to MCAO through insertion of a nylon

monofilament as early as at 30 min, peaking at 12–24 h, and falling to a few cells at day 7, while macrophages were seen at 4–6 h after occlusion. They concluded that the PMNL-associated inflammatory response contributes to delayed progressive tissue damage in focal stroke. PMNLs in the tissue may be responsible for the production of LTB_4, $-C_4$, and $-D_4$, release of large amounts of thromboxane, prostacyclin, PGE_2, and vasoconstrictors, which trigger the further chain of reactions [12]. Kochanek and Hallenbeck [25] have suggested that LTB_4 receptor binding is a potential marker of inflammatory cell infiltration in infarcted tissue.

Microglial cells which were thought to form part of the supporting structure (termed Hortega cells, mesoglia, etc.) have been shown to convert into an active form through stimulation by certain unknown mediators released from dying neurons or degenerating axons (probably IL-1). Employing a video-enhanced contrast microscope, we observed that microglia cultured on an astroglial layer begin to move by using lamellipodia [20]. The reactive microglia may release various mediators, including astrocyte growth factor which stimulates glial proliferation [7]. On the other hand, ameboid microglial cells are monocytes which migrate from blood into the tissue through the disrupted BBB.

Stage 4: Macrophages (monocytes and microglia, although it is difficult to differentiate the two types of macrophages immunochemically), which were mostly situated in the periphery, begin to increase in the ischemic core. At least several days after the insult activated monocytes reach a peak of secretory activity and release of neurotoxins [17]. Stimulation of the microglia yields soluble factors that both increase the number of astroglial cells and reduce the number of neurons. The astroglial proliferation can be blocked by incubation with an IL-1 receptor antagonist, while neurotoxic effects can be inhibited by N-methyl-D-aspartate (NMDA) receptor antagonists. Retardation of the phagocytotic activities has beneficial functional and histological effects [15]: microglia-suppressing drugs reduce the production of neurotoxic factors and improve the functional outcome from ischemic injury [14]. The amoeboid microglia plays the most active part by engulfing debris, releasing cytotoxins, killing neighboring cells, and secreting astroglial growth factors.

Stage 5: The tissue softens, liquefies, and once the debris is removed by macrophages the lesion becomes cystic. Fibroblast proliferation and therefore fibrosis is rare in an infarct. Astrogliosis is prominent in lesions with new capillaries.

Intervention

The critical role of white cells in the pathogenesis of secondary brain damage has been examined by several authors, employing animals depleted of circulating white cells by specific antineutrophil antiserum [2,18, 37], anti-CD 18 antibodies [6,27,30,41], or chemotherapeutic agents [11,23]. Table 1 summarizes the results obtained. The pretreatment protocol had beneficial effects in 5 out of 6 reports. However, administration of pharmacological agents immediately before a stroke attack is not a practical proposition in clinical cases. The effects of leukopenic or leukocyte inactivating procedures after an ischemic insult are apparently not conclusive. Other trials (lower part of Table 1), such as to scavenge oxygen free radicals with liposome-entrapped SOD[4], or to prevent immune reactions with cyclosporin A [38], chloroquine or colchicine [16] promised to be effective.

Clinical Relevance

There are several debatable issues concerning a white cell involvement in clinical ischemic stroke. Although numerous animal experiments in various settings have been performed to clarify the pathogenesis of the microvascular derangement, clinical cases of ischemic stroke differ in view of the following respects: 1) stroke patients are mostly old, whereas experimental animals are young, so that endothelial cells of the former may have somehow become predisposed (possible upregulation of adhesive molecules) due to long-standing risk factors; 2) the delicate intricacies of a long clinical course of cerebral ischemia can hardly be predicted from short experimental observations; 3) there might be species differences in the cytokines involved; and 4) specific agents such as fMLP, LPS, zymosan A, etc. are used to stimulate white cells and/ or microglia in experimental animals. Although the evolution of ischemic tissue changes resembles an inflammatory process, clinical ischemic stroke is not accompanied by 1) remarkable leukocytosis, 2) fever, 3) suppurative changes in the ischemic core, 4) correlation between the severity of ischemic stroke and leukocyte counts, or 5) beneficial effects of steroids throughout any stage of the ischemic progress. Pozzilli et al. [34] have found that leukocytosis was associated with infarct size, an impaired level of consciousness and poor clinical outcome. However, it is very difficult to rule out that the leukocytosis was due to masked infections of the respiratory tract, or urinary tract in comatose patients. Circulating white blood cells, both PMNLs and monocytes, are activated and show an increasing adhesiveness during the acute [5,19, 29,56], or even chronic phases [55] when their filter-

Table 1

Good outcome	Author	Drug	Species	Organ	Method	Duration	Parameters
Leukopenia or inactivation before ischemic insult							
Yes	Grøgaad (1989)[18]	antineutrophil antiserum	rat	forebrain	CC occlusion	1 h	CBF
Yes	Dutka (1989)[11]	mechlorethamine	dog	brain	air embolism	1 h ischemia	CBF and CSEP
No	Schott (1989)[36]	antineutrophil antiserum	dog	brain	cardiac arrest	1-2-6-12-24 h	neurologic deficit
Yes	Helps (1991)[22]	mechlorethamine	rabbit	brain	air embolism	3 h	CBF and CSEP
Yes	Bednar (1991)[2]	antineutrophil antiserum	rabbit	brain	thromboembolism	4 h	CBF and ICP
Yes	Clark (1991)[6]	anti-CD18 antibody	rabbit	spinal cord	aortic occlusion	18 h	motor function
Leukopenia or inactivation after ischemic insult							
No	Grøgaad (1989)[18]	antineutrophil antiserum 2 min after reperfusion	rat	forebrain	CC occlusion	1 h	CBF
Yes	Lindsberg (1991)[26]	anti-CD18 antibody 30 min after reperfusion	rabbit	spinal cord	aortic occlusion	18 h	motor function
Yes	Mori (1992)[29]	anti-CD18 antibody 15 min before reperfusion	baboon	brain	3 h MCAO 1 h reperfusion	3+1 h	no-reflow
No	Takeshima (1992)[40]	anti-CD18 45 min after onset of MCAO	cat	brain	1.5 h MCAO +3 h reperfusion	1.5+3 h	CBF and CSEP ischemic volume
Other trials							
Yes	Chan (1987)[4]	liposome-entrapped SOD	rat	brain	trauma	30 min-2 h-24 h	edema
Yes	Giulian (1990)[16]	chloroquine and colchicine	rabbit	spinal cord	aortic occlusion	3 days ischemia	neurological function
Yes	Shiga (1992)[37]	cyclosporin A	rat	brain	nylon thread	1 day after perfusion	edema, infarct size

CC carotid artery, *MCAO* middle cerebral artery occlusion, *CSEP* cortical sensory evoked potential, *SOD* superoxide dismutase.

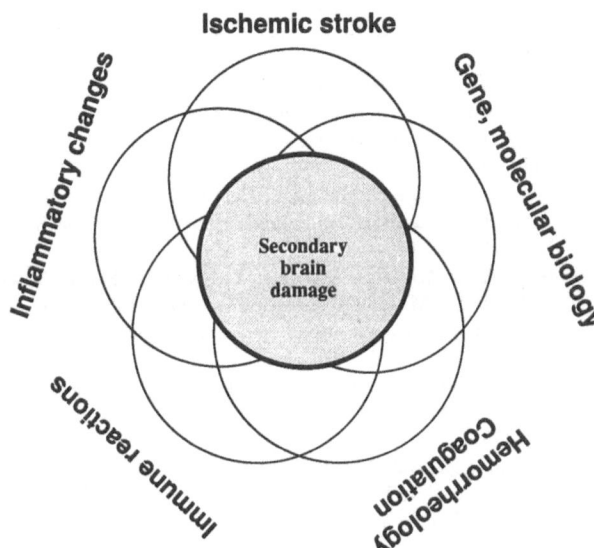

Fig. 2. Multimodality of secondary brain damage in focal cerebral ischemia

ability is examined. The order of activation is monocytes>PMNLs>lymphocytes [29]. Although Pozzilli *et al.* [34] and Wang *et al.* [57] have demonstrated leukocyte accumulation in cerebral infarct in clinical cases, a discrepancy seems to exist between actual clinical events and experimental conditions prevailing in animals. The clinical neuropathologist, Prof. Haruo Okazaki, Mayo Clinic, Rochester (personal communication), holds the view that PMNL involvement tends mostly to be limited to the perivascular space and that few PMNLs only emigrate into the ischemic tissue of the brain in ischemic stroke patients.

Our conclusions are as follows: 1) In view of the very tight cerebrovascular-parenchymal interface and dense glial sheath (*membrana limitans superficialis gliae*), the traffic of PMNLs in and out of the brain is very much restricted in the early phase of ischemia. Nevertheless, the microvascular derangement *per se* is strongly facilitated by an involvement of PMNLs adhering to endothelial cells. 2) Marked monocyte infiltration occurs with maturation of the ischemic tissue damage, which may accompany, or be accompanied by disruption of the BBB. 3) Microglia modify the process of ischemic tissue damage. The changes, thus, do not appear to comprise a spatially confined self-perpetuating process involving the ischemic tissue alone. There are various overlapping aspects (Fig. 2): ischemic stroke, genomic or molecular biological changes, inflammatory changes, immune reactions, hemorrheological disturbances, coagulation, etc. These factors occur together and are often causally interrelated. We may face the important questions

whether the inflammation-resembling processes, even though limited to the vicinity of the cerebral vessels, in fact, protect or repair the tissue rather than destroying it? Or whether the immune reaction of the microglia is harmful or beneficial for the ischemic tissue? Further, what are the mediators of the injurious mechanisms and how might they be controlled for the purpose of instituting a rational therapy?

References

1. Barone FC, Hillegass LM, Price WJ, White RF, Lee EV, Feuerstein GZ, Sarau HM, Clark RK, Griswold DE (1991) Polymorphonuclear leukocyte infiltration into cerebral focal ischemic tissue: Myeloperoxidase activity assay and histologic verification. J Neurosci Res 29: 336–345
2. Bednar MM, Raymond S, McAuliffe T, Lodge PA, Gross CE (1991) The role of neutrophils and platelets in a rabbit model of thromboembolic stroke. Stroke 22: 44–50
3. Bussolino F, Alessi D, Turello E, Camussi G (1991) Role of platelet activating factor in the adhesion process of polymorphonuclear neutrophils to endothelial cells. In: Hörl WH, Schollmeyer PJ (eds) New aspects of human polymorphonuclear leukocytes. Plenum, New York, pp 55–64
4. Chan PH, Longar S, Fishman RA (1987) Protective effects of liposome-entrapped superoxide dismutase on posttraumatic brain edema. Ann Neurol 21; 540–547
5. Ciuffetti G, Balendra R, Lennie SE, Anderson J, Lowe GDO (1989) Impaired filterability of white cells in acute cerebral infarction. B M J 298: 930–931
6. Clark WM, Madden KP, Rothlein R, Zivin JA (1991) Reduction of central nervous system ischemic injury in rabbits using leukocyte adhesion antibody treatment. Stroke 22: 877–883
7. Davis EJ, Foster TD, Thomas WE (1994) Cellular forms and function of brain microglia. Brain Res Bull 34 (1): 73–78
8. del Zoppo GJ, Garcia JH (1995) Polymorphonuclear leukocyte adhesion in cerebrovascular ischemia: Pathophysiologic implications of leukocyte adhesion. In: Granger DN, Schmid-Schönbein GW (eds) Physiology and pathophysiology of leukocyte adhesion. Oxford University Press, New York, pp 408–425
9. del Zoppo GJ, Schmid-Schönbein GW, Mori E, Copeland BR, Chang C-M (1991) Polymorphonuclear leukocytes occlude capillaries following middle cerebral artery occlusion and reperfusion in baboons. Stroke 22: 1276–1283
10. Dorovini-Zis K, Bowman PD (1992) Adhesion and migration of human polymorphonuclear leukocytes across cultured bovine brain microvessel endothelial cells. J Neuropathol Exp Neurol 51: 194–205
11. Dutka AJ, Kochanek PM, Hallenbeck JM (1989) Influence of granulocytopenia on canine cerebral ischemia induced by air embolism. Stroke 20: 390–395
12. Ernst E, Matrai A, Paulsen F (1987) Leukocyte rheology in recent stroke. Stroke 18: 59–62
13. Garcia JH, Liu KF, Yoshida Y, Lian J, Chen S, del Zoppo GJ (1994) The influx of leukocytes and platelets in an evolving brain infarct (Wistar rat). Am J Pathol 144: 189–199
14. Giulian D (1987) Ameboid microglia as effectors of inflammation in the central nervous system. Neurosci Res J 18: 155–171
15. Giulian D, Li J, Leara B, Keenen C (1994) Phagocytic microglia release cytokines and cytotoxins that regulate the survival of astrocytes and neurons in culture. Neurochem Int 25: 227–233

16. Giulian D, Robertson C (1990) Inhibition of mononuclear phagocytes reduces ischemic injury in the spinal cord. Ann Neurol 27: 33–42

17. Giulian D, Vaca K (1993) Inflammatory glia mediate delayed neuronal damage after ischemia in the central nervous system. Stroke 24 [Suppl 1]: I84–I90

18. Grøgaad B, Schürer L, Gerdin B, Arfors KE (1989) Delayed hypoperfusion after incomplete forebrain ischemia in the rat: The role of poly morphonuclear leukocytes. J Cereb Blood Flow Metab 9: 500–505

19. Grau AJ, Berger E, Sung K-LP, Schmid-Schönbein GW (1992) Granulocyte adhesion, deformability, and superoxide formation in acute stroke. Stroke 23: 33–39

20. Haapaniemi H, Tomita M, Tanahashi N, Takeda H, Yokoyama Y, Fukuuchi Y (1995) Non-amoeboid locomotion of cultured microglia obtained from newborn rat brain. Neurosci Lett 193: 121–124

21. Hallenbeck JM, Dutka AJ (1990) Background review and current concepts of reperfusion injury. Arch Neurol 47: 1245–1254

22. Hallenbeck JM, Dutka AJ, Tanishima T, Kochanek PM, Kumaroo KK, Thompson CB, Obrenovitch TP, Contreras TJ (1986) Polymorphonuclear leukocyte accumulation in brain regions with low blood flow during the early postischemic period. Stroke 17: 246–253

23. Helps SC, Gorman DF (1991) Air embolism of the brain in rabbits pretreated with mechlorethamine. Stroke 22: 351–354

24. Kobari M, Gotoh F, Tomita M, Tanahashi N, Tanaka K (1983) Vulnerability of cerebral venous flow following middle cerebral arterial occlusion in cats. In: Auer LM, Loew F (eds) The cerebral veins. Springer, Wien New York, pp 287–291

25. Kochanek PM, Hallenbeck JM (1992) Polymorphonuclear leukocytes and monocytes/macrophages in the pathogenesis of cerebral ischemia and stroke. Stroke 23: 1367–1379

26. Kulka J (1969) Injurious effects of microcirculation impairment in inflammatory disorders. In: Winters WL Jr, Brest AN (eds) The microcirculation. Thomas, Springfield, pp 174–189

27. Lindsberg PJ, Hallenbeck JM, Feuerstein G (1991) Platelet-activating factor in stroke and brain injury. Ann Neurol 30: 117–129

28. Means ED, Anderson DK (1983) Neuronophagia by leukocytes in experimental spinal cord injury. J Neuropathol Exp Neurol 42: 707–719

29. Mercuri M, Ciuffetti G, Robinson M, Toole J (1989) Blood cell rheology in acute cerebral infarction. Stroke 20: 959–962

30. Mori E, del Zoppo GJ, Chambers JD, Copeland BR, Arfors KE (1992) Inhibition of polymorphonuclear leukocyte adherence suppresses no-reflow after focal cerebral ischemia in baboons. Stroke 23: 712–718

31. Murota S, Fujita H, Morita I (1993) Involvement of adhesion molecules in vascular endothelial cell injury by oxygen radicals released from activated leukocytes. In: Tanabe T (ed) Intractable vasculitis syndromes. Hokkaido University Press, pp 115–122

32. Nakajima K, Hamanoue M, Shimojo M, Takei N, Kohsaka S (1989) Characterization of microglia isolated from a primary culture of embryonic rat brain by a simplified method. Biomed Res 10 [Suppl 3]: 411–423

33. Nawroth PP, Stern DM (1986) Modulation of endothelial cell hemostatic properties by tumor necrosis factor. Exp Med J 163: 740–745

34. Pozzilli C, Lenzi GL, Argentino C, Carolei A, Rasura M, Signore A, Bozzao L, Pozzilli p (1985) Imaging of leukocytic infiltration in human cerebral infarcts. Stroke 16 (2): 251–255

35. Rosenblum WI, El-Sabban F (1982) Influence of shear rate on platelet aggregation in cerebral microvessels. Microvasc Res 23: 311–315

36. Samuelsson B (1983) Leukotrienes: Mediators of immediate hypersensitivity reactions and inflammation. Science 220: 568–575

37. Schott R, Natale JE, Ressler SW, Burney RE, D'Alecy LG, Michigan AA (1989) Neutrophil depletion fails to improve neurologic outcome after cardiac arrest in dogs. Ann Emerg Med 18: 517–522

38. Shiga Y, Onodera H, Matsuo Y, Kogure K (1992) Cyclosporin A protects against ischemia-reperfusion injury in the brain. Brain Res 595: 145–148

39. Sirén A-L, McCarron RM, Liu Y, Spatz M, Feuerstein G, Hallenbeck JM (1993) Adhesion receptor expression and perivascular monocyte accumulation in carotid arteries and brains of hypertensive rats. In: Tomita M, Mchedlishvili G, Rosenblum WI, Heiss W-D, Fukuuchi Y (eds) Microcirculatory stasis in the brain, ICS 1031. Excerpta Medica, Amsterdam, pp 169–175

40. Suematsu M, Schmid-Schönbein GW, Chavez-Chavez RH, Yee TT, Tamatani T, Miyasaka M (1993) In vivo visualization of oxidative changes in microvessels during neutrophil activation. Am J Physiol 264: H881–H891

41. Takeshima R, Kirsch JR, Koehler RC, Gomoll AW, Traystman RJ (1992) Monoclonal leukocyte antibody does not decrease the injury of transient focal cerebral ischemia in cats. Stroke 23: 247–252

42. Tanahashi N, Fukuuchi Y, Tomita M, Kobari M, Shinohara T, Yamawaki T, Konno S, Takeda H (1993) Platelet-activating factor antagonist (TCV-309) ameliorates post-ischemic delayed hypoperfusion after 30-s cardiac arrest in cats. In: Tomita M, Mchedlishvili G, Rosenblum WI, Heiss W-D, Fukuuchi Y (eds) Microcirculatory stasis in the brain, ICS 1031. Excerpta Medica, Amsterdam, pp 203–210

43. Tanahashi N (1988) Cerebral microvascular reserve for hyperemia. In: Tomita M, Sawada T, Naritomi H, Heiss W-D (eds) Cerebral hyperemia and ischemia, ICS 764. Excerpta Medica, Amsterdam, pp 173–182

44. Tanaka K, Gotoh F, Fukuuchi Y, Amano T, Suzuki N, Uematsu D, Kawamura J, Yamawaki T, Itoh N, Obara K (1988) Stable prostacyclin analogue preventing microcirculatory derangement in experimental cerebral ischemia in cats. Stroke 19: 1267–1274

45. Tomita M (1988) Significance of cerebral blood volume. In: Tomita M, Sawada T, Naritomi H, Heiss W-D (eds) Cerebral hyperemia and ischemia, ICS 764. Excerpta Medica, Amsterdam, pp 3–31

46. Tomita M (1993) Microcirculatory stasis in the brain. In: Tomita M, Mchedlishvili G, Rosenblum WI, Heiss W-D, Fukuuchi Y (eds) Microcirculatory stasis in the brain, ICS 1031. Excerpta Medica, Amsterdam, pp 1–7

47. Tomita F, Fukuuchi Y, Tanahashi N, Kobari M, Terayama Y, Shinohara T, Konno S, Takeda H, Itoh D, Yokoyama M, Terakawa S, Haapaniemi H (1995) Activated leukocytes, endothelial cells, and effects of pentoxifylline: Observations by VEC-DIC microscopy. J Cardiovasc Pharmacol 25 [Suppl 2]: S34–S39

48. Tomita M, Fukuuchi Y, Terakawa S (1994) Differential behavior of glial and neuronal cells exposed to hypotonic solution. Acta Neurochir (Wien) [Suppl] 60: 31–33

49. Tomita M, Fukuuchi Y, Tanahashi N, Kobari M, Shinohara T, Terayama Y, Ohta K, Takeda H, Yokoyama M (1995) White cell depletion facilitates CBV recovery from ischemia following MCA occlusion in cats. Proc 18th Int Salzburg Conference, in press

50. Tomita M, Gotoh F (1992) Cascade of cell swelling (cytotoxic edema): Thermodynamic potential discharge of brain cells following membrane injury. Am J Physiol 262: H603–H610

51. Tomita M, Gotoh F, Amano T, Tanahashi N, Kobari M, Shinohara T, Mihara B (1983) Transfer function through regional cerebral cortex evaluated by a photoelectric method. Am J Physiol 245: H385–H398

52. Tomita M, Gotoh F, Amano T, Tanahashi N, Tanaka K (1980) Low perfusion hyperemia following middle cerebral arterial occlusion in cats of different age groups. Stroke 11: 629–636

53. Tomita M, Gotoh F, Sato T, Amano T, Tanahashi N, Tanaka K, Yamamoto M (1978) Photoelectric method for estimating hemodynamic changes in regional cerebral tissue. Am J Physiol 235: H56–H63

54. Turcani P, Gotoh F, Tomita M, Tanahashi N, Kobari M, Terayama Y, Mihara B, Ohta K (1987) Role of platelets and leukocytes in the development of low perfusion hyperemia in the cerebral ischemic area of cats. In: Meyer JS, Lechner H, Reivich M, Ott EO (eds) Cerebral vascular disease 6. Excerpta Medica, Amsterdam, pp 285–289

55. Vermes I, Strik F (1988) Altered leukocyte rheology in patients with chronic cerebrovascular disease. Stroke 19: 631–633

56. Violi F, Rasura M, Alessandri C, Intiso D, Germani M, Servi M, Balsano F (1988) Leukocyte response in patients suffering from acute stroke. Stroke 19: 1283–1284

57. Wang PY, Kao CH, Mui MY, Wang SJ (1993) Leukocytic infiltration in acute hemispheric ischemic stroke. Stroke 24: 236–240

58. Yokoyama M, Fukuuchi Y, Tomita M, Tanahashi N, Kobari M, Konno S, Takeda H, Ito D, Terakawa S (1994) PMNL activation: Observations by VEC microscopy (Abstract). In: Tsuchiya M, Asano M, Ohhashi T (eds) Microcirculation Annual 1994. Nihon-Igakukan, Tokyo, pp 165–166

Correspondence: Minoru Tomita, M.D., Department of Neurology, School of Medicine, Keio University, 35 Shinanomachi, Shinjuku-ku, Tokyo 160, Japan.

Acta Neurochir (1996) [Suppl] 66: 40–43
© Springer-Verlag 1996

Inflammation of the Brain after Ischemia

K. Kogure[1], **Y. Yamasaki**[2], **Y. Matsuo**[3], **H. Kato**[4], and **H. Onodera**[4]

[1]Department of Pharmacology, Tokyo Medical College and Institute Neuropathology, Saitama, [2]Hanno Research Center, Taiho Pharmaceutical Co. Ltd., [3]Developmental Research Laboratory, Shionogi and Co. Ltd., and [4]Department of Neurology, Tohoku University School of Medicine, Japan

Summary

Cytokines which promote emigration of leukocytes from the vascular lumen into the injured brain tissue are produced at the site of incipient cerebral infarction. The blood-borne invaders then accelerate the decomposition of brain cells by their toxic by-products, phagocytic action, and by the immune reaction. Recently accumulated data in our laboratories and other research facilities show that depleting the amount of circulating leukocytes or administering anti-inflammatory chemicals such as cytokine blocking agents, anti-adhesion molecule antibodies, and immunosuppressants effectively minimize the size of ischemia induced cerebral infarction. Based on the fact that leukocyte invasion of the affected brain tissue occurs 6 to 24 hours after onset of ischemia, administration of an anti-inflammatory therapy may widen the therapeutic window against stroke.

Keywords: Cerebral infarction; stress protein; cytokines; inflammation of the brain.

Introduction

Ischemic brain cell death has been generally described as "death by energy failure" leading to pan-necrosis. The temporal evolution of ischemic infarction can take minutes to hours, and the focal brain damage is almost always associated with secondary brain cell injury which takes place hours to days after the initial ictus.

In this regard, however, damage and subsequent cell reaction to ischemia vary according to the severity and length of ischemia. If and ischemic insult is mild or of short duration, only selectively vulnerable neurons are damaged. As the ischemic period is prolonged, the selectiveness of cell damage is lost, and tissue damage progresses in a more general fashion. In the case of severe focal ischemia, the center becomes pan-necrotic, and this is termed infarction.

Intra-Ischemic Events During a Lack of Energy

After inhibition of energy production by ischemia, the immediate complication is a stalling of the ion-pumping ATPases with subsequent breakdown of transmembraneous ionic gradients. Failure of Na, K-ATPase activity, the most noticeable immediate change, allows intracellular potassium ions to leak into the extracellular space and sodium ions to accumulate intracellularly. This produces membrane depolarization, leading to influx of Ca^{2+} through activation of voltage dependent Ca^{2+} channels. Intracellular Ca^{2+} accumulation in presynaptic neurons causes a nonspecific, nonphysiological release of neurotransmitters from presynaptic terminals, which in turn facilitates agonist-(receptor-) mediated hydrolysis of inositol phospholipids through activation of phospholipase C. As a consequence, triphosphoinositol is cleaved into 1, 2-diacylglycerol (DG) and inositol 1, 4, 5-triphosphate (IP_3). In the ischemic brain, DG is decomposed by di- and monoacylglycerol lipases, resulting in the liberation of free fatty acids. IP_3 on the other hand, increases the cytoplasmic Ca^{2+} concentration by releasing Ca^{2+}-ions from intracellular Ca^{2+} stores.

In postsynaptic neurons, in addition to the opening of voltage-dependent Ca^{2+} channels, receptor-operated Ca^{2+} channels are activated by neurotransmitters such as glutamate and aspartate. This facilitates again a marked increase in the concentration of cytosolic Ca^{2+} combined with the mobilization of Ca^{2+} by IP_3 and IP_4. This increase in cytosolic Ca^{2+}-ions induces nonspecific activation of phospholipase A_2 and of other phospholipases, which subsequently accelerate decomposition of membraneous structural phospholipids, such as phosphatidyl choline (PC) and phosphatidyl ethanolamine (PE). Liberation of palmitic- and docosahexaenic acids, which indicates a decomposition of PC and PE starts 2 to 3 min after the onset of ischemia [2]. On the other hand, there have been reports that prolonged ischemia, as long as 30 min decomposes 16% of membrane PE [14], and that 37% of free arachidonic acid accumulated during 30 min of ischemia is released from membrane PC [5].

Post-Ischemic Events Following Restoration of Energy

Needless to say that brain tissue cannot survive under continuous ischemia in a state of permanent energy failure. If, however, effective blood flow is restored to the tissue before affected brain cells have passed a point of no-return toward necrosis, in the tissue a stereotypical recovery process commences. Intermediate metabolites derived from the above mentioned decomposition of membraneous phospholipids serve as second messengers and initiate life saving cascade reactions, both in neurons as well as in surrounding glia cells.

Restoration of blood flow in post-ischemic brain tissue might be compared with the supply of serum to incubated brain cells in serum-free incubation media. Similar to the serum supply, post-ischemic reperfusion re-energizes affected brain cells. Through phosphorylation of regulatory enzymes neurons and glial cells start to express immediate early genes, even though general protein synthesis in these brain cells is markedly suppressed due to a disordered amino acid transport across the plasma membrane and degradation of ribosomes from polysomes to monosomes.

Thereafter, some of the corresponding immediate early gene products, the proteins contribute to the formation of transcription factors, such as nerve-growth-factor-inducible protein A (NGFI-A), which contains a zinc-finger domain, fos/jun families which contain a leucine zipper domain, c-myc (leucine zipper and helix-loop-helix), and NGFI-B (steroid hormone receptor transcription factor). It is noteworthy that even though the expression of heat shock protein 70 mRNA (HSP 70 mRNA) has been hitherto unknown, it becomes clear that induction of the zinc-finger gene proceeds through expression of HSP 70 mRNA. A possible assumption based on this is that these transcription factors may play important roles in the ex-

pression of heat shock proteins and perhaps in expression of other types of stress proteins including amyloid precursor protein (APP) 751 and 770 mRNA. APP 751 and 770 are known to contain the Kunitz-type protease inhibitor domain, which has been recognized as protease nexin II, a strong chymotrypsin inhibitor [10]. Superoxide dismutase, catalase, ornithin decarboxylase, and other inducible proteins are also termed stress proteins, even though their actual roles have not yet been clarified. Many members of the heat shock protein family (from 8 kDa ubiquitin to 110 kDa HSP) are known as molecular chaperones for degradative, unfolded, or newly synthesized nascent polypeptides.

The above-mentioned cascading reactions, namely expression of immediate early genes, induction of the corresponding formation of transcription factors, and of stress proteins may increase resistance of the affected brain cells to the provocated ischemic changes. Once such conditions are established the cell acquires a certain tolerance (reduced susceptibility) towards secondary ischemia. Mechanisms underlying this phenomenon may also involve down-regulation of intracellular receptors of the signal transduction system beside de novo synthesis of stress proteins.

However, if this organized response does not suffice for the brain cells to escape impending cell death, neurons and glial cells synthesize then a variety of trophic factors and up-regulate corresponding receptors on the surface of the target cell membrane. In this fashion, activated astrocytes, and perhaps microglial cells attached to or in the vicinity of troubled neurons support their recovery processes.

Mechanism Underlying Ischemia-Reperfusion Injury in Focal Lesions

If the above-mentioned lifesaving steps in a focal lesion fail to maintain viability of the affected brain tissue, additional defense is sought through the induction of cytokines and establishment of so-called cytokine networks. For instance, interleukin-I (IL-1) facilitates induction of SOD in neurons, proliferation of astroglia and induces intercellular adhesion molecule-I (ICAM-I) on the endothelial surface. Tumor necrosis factor-α (TNF-α) and interferon-γ (IFN-γ) secreted by activated T cells induce Class I and II major histocompatibility complex (MHC) antigens on the surface of neurons and astroglia [4].

Particularly, TNF-α and IL-1 β are believed to be powerful up-regulators of leukocyte-endothelium adhesion molecules (CD-11a and CD18). The cytokine-induced neutrophil chemoattractant (CINA)/IL-8 is suggested to play a critical role in invasion of damaged brain tissue by neutrophils [11]. Such post-ischemic inflammation of the brain consists of primary invasion by neutrophils which is then followed by immigration of monocytes.

Recent data accumulated in our laboratories and other institutes strongly suggest that together with activated phagocytic microglia, emigrated neutrophils, blood-borne transformed macrophages, and immunoreactive lymphocytes may actually kill brain cells surviving the primary ischemic insult. If this happens in peripheral tissues, the resultant necrotic mass is eventually replaced by vital tissue through proliferation of surviving cells. Inflammation can there be considered as a process of healing. In contrast, neurons in the brain do not have an ability to proliferate. Thus, the result of the inflammatory scavenging process is either liquefaction of the debris or a fibrosing scar. Presumably the brain might be considered as a newcomer in front of the evolutionary well-established defense line. It could be recognized by the defense system as being of different nature once the blood-brain barrier is stripped off due to endothelial disruption and astrocytic dysfunction.

Summarizing the above-mentioned evolving processes, one may propose the following reaction sequence:

I. Intra-Ischemic Reaction

1) energy failure; 2) failure of ion pumps; 3) breakdown of ionic and water homeostasis; 4) decomposition of the plasma membrane and the micro-organella; 5) autolysis of cells and pan-necrosis of the lesion (development of infarction).

II. Post-Ischemic Reaction
with Restored Energy Metabolism

1. Reaction confined within the post-ischemic tissue with higher energy state

1) expression of immediate early genes and induction of corresponding proteins; 2) formation of transcription factors; 3) induction of molecular chaperones and other stress proteins; 4) induction of trophic factors and up-regulation of corresponding receptors.

2. Reaction spreading out from tissue with a low energy state

5) induction of cytokines and up-regulation of receptors; 6) formation of cytokine networks; 7) leukocyte invasion and inflammation; 8) exaggerated immune reaction; 9) pan-necrosis of the lesion (resultant infarction).

Identification of the Point of No-Return

It has been generally considered that the life of cells is lost following decomposition of the plasma membrane. Viability of the cell requires that the internal matrix of the cell is physically and chemically isolated from the external environment to maintain its own homeostasis utilizing the plasma membrane as a barrier. But, one may entertain different views with regard to various experimental data available as follows;

If the degree of intra-ischemic injury is not so serious making possible improvement of the energy state by a restored blood flow, intermediate metabolites of membraneous phospholipids can act as second messengers which induce a programed process of reconstruction allowing the cell to recover its structural and functional integrity.

If the degree of ischemia is serious, not only neurons but also glial cells participate in the recovery process. If, however, the ischemia-induced brain cell injury is even more severe, the damaged tissue is subjected to an inflammatory reaction, which seems to cause a final destruction of brain cells. At this stage, the neurons must have passed the point of no-return.

Abe [1] and Aoki [2] reported recently that transient cerebral ischemia may cause dysfunction of the mitochondrial shuttle system in neurons, the axonal terminal and dendrites. Once this happens, mitochondria may fail to obtain newly synthesized nuclear encoded mitochondrial protein.

Thus, the cell may suffer from a secondary and irreversible energy failure. Occurrence of this phenomenon is also suspicious to represent a point of no-return. From a logistic point of view the above-mentioned cytokine synthesis can occur only, if the cell is capable to transcribe messages encoded in DNA, and ribosomes are capable to conduct the translational procedure. For this reason, we assume that the critical time point discussed above should be earlier than the secondary precipitation of the energy state.

Prospective Drugs Against
an Imminent Cerebral Infarction

Based on the above sequence of events, specific strategies may be developed for the pharmacological treatment of acute stroke. Prospective methods are:

1) depletion or inactivation of circulating leukocytes; 2) blocking of adhesion molecules; 3) blocking of the release of cytokines and of their effects; 4) administration of immunosuppressants to attenuate exaggerated immunoreactions.

We have tested the above propositions with the following results:

1) Administration of an anti-neutrophil monoclonal antibody (RP-3) [8], x-ray irradiation of rats [13], and pretreatment of animals with colchicine [8] were markedly suppressing formation of post-ischemic brain edema and/or minimizing the size of infarction. 2) Anti-adhesion molecule antibodies, named 1A-29 (anti-ICAM-1 antibody), both WT-1 and WT-3 (anti-CD 11a and CD 18 antibody, respectively) were effectively suppressing post-ischemic brain damage [6]. 3) PAF antagonist (Y-24180, unpublished data by Matsuo et al.), IL-1 blocking agent (ZnPP), anti-IL-1 antibody, and IL-1 release inhibitor, were blocking expansion of post-ischemic brain injury [12]. 4) Immunosuppressants such as cyclosporin A [9] were minimizing post-ischemic tissue damage.

Professor Scheinberg once expressed his dissatisfied feeling on recently available clinical drugs against acute stroke and commented that "Treatment of acute stroke is still disappointing despite the development of many promising pharmacological strategies in experimental animals. An important part of the reason may be that opening of the window of therapeutic opportunity is much shorter than the usual entry time of patients in most clinical trials. This logistic problem merits serious attention" [7].

From his follower's experimental data as reviewed here, it is concluded that it may be possible to design useful drugs against ischemic stroke. The use of anti-inflammatory agents may widen the window of therapeutic opportunity.

References

1. Abe K, Kawagoe J, Aoki M, Kogure K (1993) Disturbance of mitochondrial DNA expression in gerbil hippocampus after transient forebrain ischemia. Mol Brain Res 19: 69–75
2. Abe K, Kogure K, Yamamoto H, Imazawa M, Miyamoto K (1987) Mechanism of arachidonic acid liberation during ischemia in gerbil cerebral cortex. J Neurochem 48: 503–509
3. Aoki M, Abe K, Yoshida T, Hattori A, Kogure K, Itoyama Y (1995) Early immunohistochemical changes of microtubule based motor proteins in gerbil hippocampus after transient ischemia. Brain Res 669: 189–196
4. Benveniste EN (1992) Inflammatory cytokines within the central nervous system: Sources, function and mechanism of action. Am J Physiol 263: C1–16
5. Marion J, Wolfe LS (1979) Origin of the arachidonic acid released post-mortem in rat forebrain. Biochem Biophys Acta 574: 25–32
6. Matsuo Y, Onodera H, Shiga Y, Shozuhara H, Ninomiya M, Kihara T, Tamatoni T, Miyasaka M, Kogure K (1994) Role of cell adhesion molecules in brain injury after transient middle cerebral artery occlusion in the rat. Brain Res 656: 344–352
7. Scheinberg P (1994) Stroke: the way things really are. Stroke 25: 1290–1294
8. Shiga Y, Onodera H, Kogure K, Yamasaki Y, Shozuhara H, Sendo F (1991) Neutrophil as a mediator of ischemic edema formation in the brain. Neurosci Lett 125: 110–112
9. Shiga Y, Onodera H, Matsumoto Y, Matsuo Y, Kogure K (1992) Cyclosporin A protects against ischemia-reperfusion injury in the rat brain. Brain Res 595: 145–148
10. William E, Nostrand U, Wagner SL, Suzuki M, Choi BH, Farrow JS, Geddes JW, Cotman CW, Cunningham DD (1980) Protease nexin-II a potent antichymotrypsin, shows identity to amyloid β-protein precursor. Nature 341: 546–549
11. Yamasaki Y, Matsuo Y, Matsuura N, Onodera H, Itoyama Y, Kogure K (1995) Transient increase of cytokine-induced neutrophil chemoattractant, a member of the interleukin-8 family, in ischemic brain areas after focal ischemia in rats. Stroke 26: 318–323
12. Yamasaki Y, Suzuki T, Yamaya H, Matsuura N, Onodera H, Kogure K (1992) Possible involvement of interleukin-1 in ischemic brain edema formation. Neurosci Lett 142: 45–47
13. Yamasaki Y, Yamaya H, Watanabe M, Matsuura N, Onodera H, Kogure K (1991) Lymphocytes play a critical role in the ischemic brain edema formation. J Cereb Blood Flow Met 13 [Supppl] 1: S 113
14. Yoshida S, Inoh S, Asano T, Sano K, Kubota M, Shimazaki H, Ueta N (1980) Effect of transient ischemia on free fatty acids and phospholipids in the gerbil brain: Lipid peroxidation as a possible cause of postischemic injury. J Neurosurg 53: 323–331

Correspondence: Kyuya Kogure, M.D., Department of Pharmacology, Tokyo Medical College and Institute Neuropathology. Saitama, Japan.

Acta Neurochir (1996) [Suppl] 66: 44–49

Three-Dimensional Metabolic and Hemodynamic Imaging of the Normal and Ischemic Rat Brain

M. D. Ginsberg, T. Back*, and W. Zhao

Department of Neurology, Cerebral Vascular Disease Research Center, University of Miami School of Medicine, Miami, FL, U.S.A.

Summary

Unique insights into the topography of local metabolism/blood flow interrelationships in focal cerebral ischemia have been afforded by the recent development of powerful image-processing techniques permitting three-dimensional (3D) autoradiographic image-averaging and analysis of replicate studies by a novel method termed "disparity analysis". This method, based upon a linear affine transformation model, directly estimates scaling, translation and rotation parameters simultaneously. The method was validated in awake Wister rats studied for local cerebral glucose metabolism (lCMRgl) with ^{14}C-2-deoxyglucose. Brains were subserially sectioned, aligned by disparity analysis, and mapped into a common template so as to generate aggregate 3D data sets of the mean and standard deviation of the entire series (n=9). Internal anatomic architecture was faithfully represented in the average image, and Fourier analysis revealed satisfactory retention of low-frequency information. The method was then applied to study metabolism/blood flow relationships in the acute focal ischemic penumbra of Sprague-Dawley rats subjected to distal photothrombotic middle cerebral artery (MCA) occlusion, coupled with permanent ipsilateral and 1 h contralateral common carotid artery occlusions. Matched series were studied for lCBF at 1.5 h and for lCMRgl at 1.25-2 h post-occlusion. The averaged lCBF image revealed the ischemic penumbra (defined as lCBF 20–40 % of control) to form a "shell" around the cortical ischemic core and a confluent aggregate at the anterior and posterior poles of the core-zone. lCMRgl in the penumbra was heterogeneous, ranging from near-normal to markedly increased. An average lCMRgl/lCBF *ratio* data set revealed marked metabolism-flow uncoupling in penumbral pixels, averaging nearly five-fold above control ratio values. Sustained deflections of the DC potential were recorded in the penumbra, the site of marked uncoupling. This analysis defined for the first time the 3D topography of the ischemic penumbra and substantiated marked metabolism/flow dissociation, which is believed to be a metabolic consequence of the energy demand imposed by repeated peri-infarct depolarizations.

Keywords: Image-processing; local cerebral blood flow; glucose metabolism; disparity analysis.

Introduction

Traditionally, autoradiographic studies of local cerebral glucose utilization (lCMRgl) or blood flow (lCBF) have been carried out on representative coronal sections of individual animal brains. Access to an *entire three-dimensional (3D) autoradiographic image data set*, however, would in principle offer several advantages: the ability to view the brain from multiple planes or sections; and, more importantly, the potential to *average* data from replicate series, which would greatly simplify region-of-interest analysis and would provide the capability for arithmetic manipulations of data sets derived from different series. The development of 3D autoradiography has been limited by methods available for the alignment (registration) of sequential sections. Previous registration methods have included the principle-axes method [6], which is efficient but unsatisfactory if images tend to be round or if bilateral image symmetry is lacking (as in damaged sections). A second approach, cross-correlation, utilizes similarities of spatial distributions of gray-level values [3]. A third approach, feature-matching, necessitates selection of the same feature points from two images and becomes less robust if fewer common features can be extracted [9].

We developed a novel approach for aligning autoradiographic images that overcomes most difficulties experienced by previous methods [17,18]. This method, termed "disparity analysis," is based upon a linear affine transformation model to analyze point-to-point

---

* Present address: Neurologische Universtätsklinik der Charité, Humboldt-Universität, Berlin, Federal Republic of Germany

disparities in two images and is a direct method that estimates scaling, translation and rotation parameters simultaneously without second-step transformations. The method is general and flexible, utilizing the same basic principle to deal with different situations such as damaged or asymmetric sections. The theory of this method has been previously presented [18], and the method has been recently validated in a series of studies in normal rats, described below [17].

A major challenge in studying the pathophysiology of cerebral ischemia is to achieve a precise metabolic and hemodynamic appreciation of the ischemic penumbra – a temporally and spatially dynamic region of intermediate lCBF [4,7]. Success in developing 3D autoradiographic image-averaging has allowed us to investigate this problem precisely, using a highly serviceable photothrombotic model of distal middle cerebral artery (MCA) occlusion. Below, we describe as well the major findings of that study [1].

Theory of Image Registration by Disparity Analysis

As proposed by Henderson et al. [5], let $f(i, j)$ and $g(k, l)$ be two digital images. Let $C = \{(i, j; k, l)\}$ be the geometric location (i, j) in f, corresponding to the geometric location (k, l) in g. $i, j; k, l$ belong to set R. We say that T geometrically registers f with respect to g if

1) $T : R \times R \rightarrow R \times R$, and
2) $T(k, l) = (i, j)$ for every $(i, j; k, l)$ in C.

T is called the registration transform (or function) of f with g. The set C is called the control point set, where every four-tuple defines one control point pair: (i, j)–(k, l). In order to register f with respect to g, we define

$$f'(i, j) = f(T(i, j)) \tag{1}$$

as the registered image of f with g. This intermediate step of computing f' is performed in reverse for registration since T produces a value at each pixel of f' by going back to a pixel (or neighborhood) in f [17,18].

Let $\underline{x}^T = (x, y)$ be a point on the boundary of a 2D coronal section and $\underline{x}'^T = (x', y')$ be the corresponding point in the second image. We assume an affine transformation relationship between the two points,

$$\underline{x}' = \underline{A}\underline{x} + \underline{c} + \underline{e}(\underline{x}), \tag{2}$$

where $\underline{A}$ is a 2×2 affine matrix, and where $\underline{e}(\underline{x})$ represents noise and/or minor local differences of the two boundaries. The disparity vector $\underline{d}(\underline{x})$ is simply

$$\underline{d}(\underline{x}) = \underline{x}' - \underline{x} = \underline{B}\underline{x} + \underline{c} + \underline{e}(\underline{x}), \tag{3}$$

where $\underline{B} = \underline{A} - \underline{I}$ and $\underline{I}$ is the 2×2 identity matrix.

The physical meaning of this linear affine model is as follows: 2D translation is represented by the vector $\underline{c}$. The matrix $\underline{A}$ contains information including 2D rotation, 2D angular deformation (shearing), and height and width changes. It has been shown [15,16] that 3D rotation or tilting of a planar patch results in linear shape changes that can be represented by a 2D matrix operation. Indeed, the affine matrix $\underline{A}$ can be uniquely decomposed:

$$\underline{A} = \underline{A}_l \underline{A}_a \underline{A}_r \begin{pmatrix} l_x & 0 \\ 0 & l_y \end{pmatrix} \begin{pmatrix} \cos \alpha & \sin \alpha \\ \sin \alpha & \cos \alpha \end{pmatrix}$$

$$\begin{pmatrix} \cos \theta & -\sin \theta \\ \sin \theta & \cos \theta \end{pmatrix} \tag{4}$$

With simple algebraic and trigonometric manipulations, we obtain (16),

$$l_x = \sqrt{a_{11}^2 + a_{12}^2},$$

$$l_y = \sqrt{a_{21}^2 + a_{22}^2},$$

$$\alpha = 0.5 \left(\arctan\left(\frac{a_{21}}{a_{22}}\right) + \arctan\left(\frac{a_{12}}{a_{11}}\right) \right),$$

$$\theta = 0.5 \left(\arctan\left(\frac{a_{21}}{a_{22}}\right) - \arctan\left(\frac{a_{12}}{a_{11}}\right) \right). \tag{5}$$

In other words, for a given $\underline{A}$, the 2D rotation angle θ, the deformation angle α, and the height and width changes (scale factors), l_x and l_y, can be computed uniquely from the matrix elements.

Consider N points on the boundary of a coronal section $\underline{x}_i$, $i = 1, 2, ..., N$. If the corresponding boundary points $\underline{x}'_i$, $i = 1, 2, ..., N$, in the next image were known, the point-to-point matching problem would be solved. The image point $\underline{x}_i$ differs from its corresponding point $\underline{x}'_i$ by a disparity vector, $\underline{d}_i = \underline{d}(\underline{x}_i) = \underline{x}'_i - \underline{x}_i$. Thus, disparity analysis is an automated method to locate a large number of pairs of corresponding points. The 2D vector $\underline{d}_i$ can be decomposed into a component tangential to the boundary in the first image and a component perpendicular to the tangent. The tangential component cannot be measured without knowing $\underline{d}_i$. For the perpendicular component, let us superimpose the second boundary on the first one and draw a perpendicular to the tangent of the first boundary at point $\underline{x}_i$ until it intersects the second boundary. The difference between the intersection point and $\underline{x}_i$ is then the measured perpendicular component. Let $\underline{n}_i$ be the normal vector in the direction of the perpendicular component and v_i

be the magnitude of the component. Then, the magnitude of perpendicular component is the projection of disparity vector onto the normal vector,

$$v_i = \underline{n}_i^T \underline{d}_i. \tag{6}$$

The disparity vector $\underline{d}_i$ cannot be recovered from the measured v_i alone. However, if we substitute Eq. (3) into (6), we can formulate a mean-square error problem with respect to a set of optional elements in $\underline{A}$ and $\underline{c}$.

$$\varepsilon = \sum_{i=1}^{N} (\underline{n}_i^T (\underline{B}\underline{x}_i + \underline{c}) - v_i)^2 \tag{7}$$

where ε is due to the differences between the projection of the computed disparity vector onto the normal vector and the measured perpendicular components. Detailed derivations of a closed-form solution of $\underline{A}$ and $\underline{c}$ can be found in [18].

Materials and Methods – Validation Studies

Local cerebral glucose utilization (lCMRgl) was measured in 9 fasted awake male Wister rats, weighing 240–350 g [17]. Animals were initially anesthetized with halothane by face mask for insertion of femoral catheters to monitor arterial blood pressure and to sample arterial blood gases. Rectal temperature was maintained at 37 °C. Animals were then encased in a plaster body cast allowing partial movements of forelimbs and hindlimbs; the cast was attached to a lead brick, and animals were allowed to awaken in a quiet, dimly lit room. They were studied 2 h after recovering from anesthesia. Physiological values were within normal limits, with $PCO_2 = 34.0 \pm 3.5$ mm Hg, arterial pressure = 110 ± 5 mm Hg, and plasma glucose = 142 ± 23 mg/dl (means $\pm$ SD).

lCMRgl was measured with ^{14}C-2-deoxyglucose in the usual manner [13]. Brains were rapidly removed and frozen by gradual immersion in liquid nitrogen; they were subsequently transferred to a cryostat at –20 °C, carefully aligned with respect to their antero-posterior axes, and sectioned subserially at 100 µm intervals. Sections and calibrated standards were placed on Kodak Hyperfilm βmax film for a 10-day exposure. Films were then scanned at 100 µm resolution on a rotating-drum densitometer (Optronics) interfaced to an image-processor and Micro-VAX computer system. To achieve 3D image-alignment, the disparity analysis method was first applied to align the approximately 200 coronal forebrain sections from each brain. Each of the 9 resulting 3D image data sets could then be inspected as a 3D stack or as coronal sections (Fig. 1). The 9 image sets were then co-registered with one another with respect to their longitudinal axis, using a standard reference level (bregma + 0.7 mm, [11]). The disparity analysis method was then reapplied to co-align respective coronal data sets at each level. This yielded an *average* 3D image data set as well as a 3D data set of the standard deviation. To achieve this result, 1 of the 9 brains was designated the "template,"

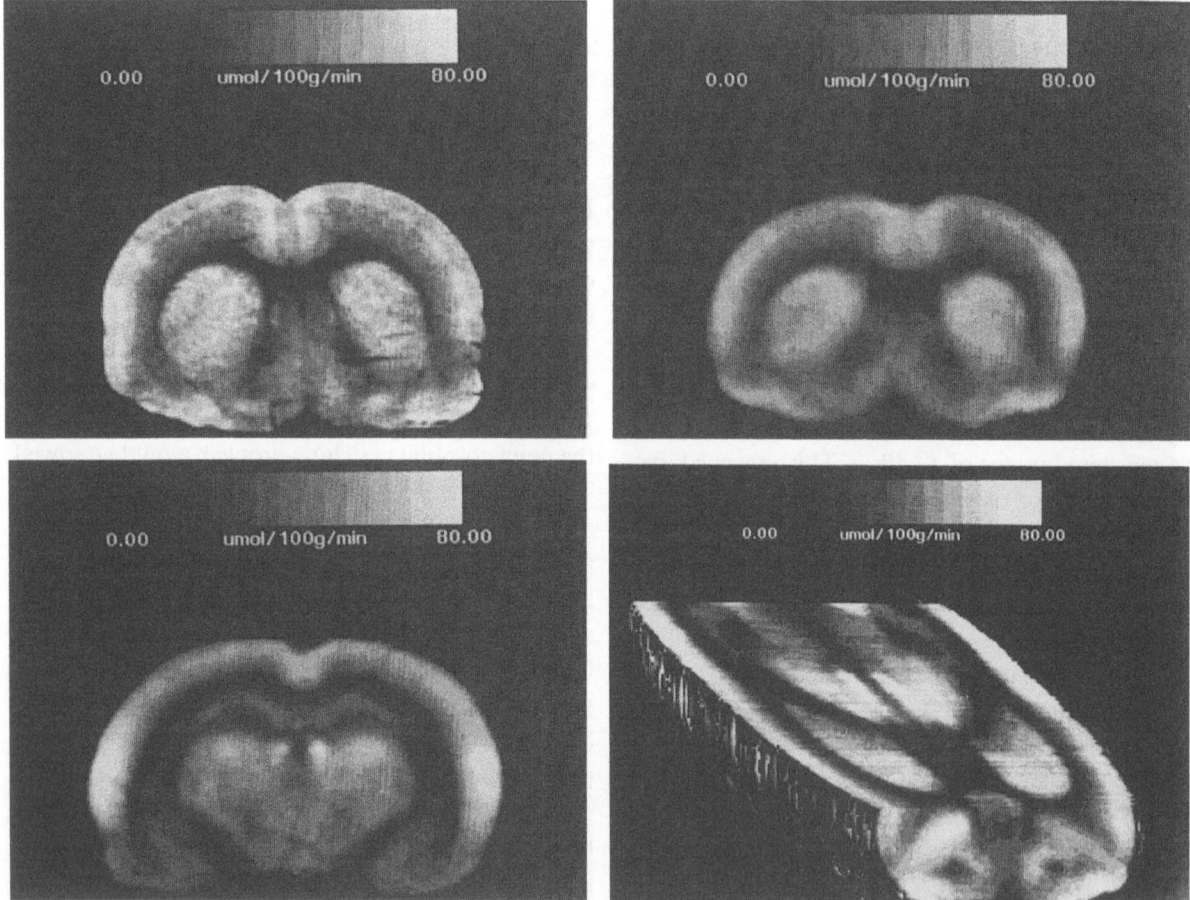

Fig. 1. ^{14}C-2-deoxyglucose autoradiographic studies of local cerebral glucose utilization (lCMRgl) in normal awake Wistar rats. Upper left: Digitized coronal section derived from an *individual* brain. Upper right: Coronal section from the *averaged* image data set at the same level. Lower left: Coronal section from averaged data set at level of mid-thalamus. Lower right: Horizontally sectioned average 3D data set

and the other 8 were mapped into its external contour. Fourier analysis of line-scans of individual and average coronal images was undertaken in order to assess the fidelity of the average image data set with respect to its internal architecture, when compared to individual autoradiograms.

Results and Discussion – Validation Studies

Image Deformation and Averaging

Figure 1 depicts a coronal section from a typical individual brain at the level of mid-striatum, and compares it to the section from the *average* data set at the same level. The average image retains the major features apparent in the individual section, with a slightly "defocused" appearance. Figure 1 also displays the 3D image data set sectioned in the horizontal plane; the internal anatomic features (caudoputamen, thalamus) of this averaged brain are smooth, continuous and readily identifiable.

Feature-Comparison Studies

For this analysis, we constructed lines on 4 coronal sections which traversed defined anatomic structures (e.g., caudoputamen, thalamus, dorsal hippocampus, medial geniculate body, auditory cortex). These lines were scanned on the template brain, on the 8 other individual brains following their deformation to the template, and on the average image data set. Mean correlation coefficient for the deformed sections vs. template sections was 0.93 ± 0.01. Line-scans through the averaged autoradiograms were very similar to the template but appeared smoother, and the troughs tended to be somewhat truncated. Fourier analysis revealed that spectra representing the averaged sections retained most of their low-frequency information but had somewhat less high-frequency information compared to individual brains. Region-of-interest lCMRgl measurements were carried out on representative structures of the averaged brains at four coronal levels and were compared to ROI measurements taken from individual brains and then averaged. The mean percent difference between the two sets of readings for 24 structures was $3.4 \pm 5.0 \%$, and the correlation coefficient between the two data sets was 0.99 ($p < 0.0001$) [17].

To summarize, 3D autoradiographic image-averaging of a replicate series yields accurate and highly serviceable data. Advantages of this approach include: 1) the more ready identification of group trends in the average image than would be apparent from inspection of individual data sets; 2) a vast simplification of region-of-interest analysis since ROI analysis can be performed directly on the average and standard-deviation data sets; 3) this method allows volumes-of-interest (defined with respect to certain threshold levels of metabolism or blood flow) to be readily computed from the averaged 3D data set; 4) an atlas-based ROI strategy becomes feasible as an atlas overlay can be mapped into the common template; 5) the ability to map several series of brains into a common 3D space provides the means for rigorous comparison of two or more animal series studied under varying conditions or by different radiotracer strategies. An example of the latter forms the basis of the second part of this report [1].

Materials and Methods – Focal Ischemia Studies

These studies were carried out in fasted male Sprague-Dawley rats weighing 250–265 g, which were anaesthetized with halothane, endotracheally intubated, immobilized with pancuronium and artificially ventilated. Femoral vessels were catheterized. Rectal and ipsilateral temporalis muscle temperatures were maintained at 37 °C by feedback-controlled heating lamps. The common carotid arteries (CCA) were exposed bilaterally. A surgical procedure previously described [10,14] was used to expose the right distal middle cerebral artery above the rhinal fissure via a small temporal craniotomy, leaving the dura intact. The right MCA was photothrombotically occluded by irradiating three of its distal branching points as previously described [10], utilizing an argon laser-activated dye laser tuned to 562 nm at power of 20 mW; the single laser beam was split into three beams, which were positioned individually at the designated branch points. The photosensitizing dye rose bengal (15 mg/ml, 1.34 ml/kg) was administered i.v. prior to MCA irradiation. Five min after onset of irradiation, the snare ligatures were tightened around both CCAs. The right CCA remained permanently occluded, and the contralateral CCA occlusion was released after 1h.

Rats were randomly assigned to two groups (n=7 each). In one group, local cerebral blood flow (lCBF) was measured with ^{14}C-iodoantipyrine injected 1.5 after induction of ischemia [12]. In the second group, local cerebral glucose utilization (lCMRgl) was measured with ^{14}C-2-deoxyglucose, administered 1.25 h post-MCA occlusion, with a 45-min study period. Brains were removed and processed for autoradiography as described above (see Validation Studies). 3D autoradiographic image data sets were constructed for each animal. After alignment of individual lCBF or lCMRgl images, corresponding coronal sections from all brains of the same series were registered with one another with respect to a common coronal reference level. One brain of each series served as a template, and the corresponding sections of the other brains were mapped into its contour by means of the averaging procedure described above under Validation Studies. In this manner, quantitative 3D reconstructions of *averaged* lCBF and lCMR data sets were computed and could be displayed. Standard deviation data sets were also computed. An average CMRgl/CBF *ratio* data set was then calculated essentially by dividing every individual lCMRgl study by every individual lCBF study and subsequently double-averaging (details presented in [1]).

A computer thresholding tool allowed us to produce selective 3D constructions depicting only those pixels lying within predefined thresholds for lCMFgl, lCBF or the lCMRgl/lCBF ratio. We elected specifically to analyze those pixels of the 3D lCBF data set having blood flow in the moderately reduced range (defined as 20–40 % of contralateral values) to permit the study of metabolism-flow uncoupling in the ischemic penumbra.

Results and Discussion – Focal Ischemia Studies

Physiological variables were within the normal range in both groups. Mean CBF of the non-ischemic hemisphere was 1.17 ± 0.48 ml/g/min [1], and no focal areas of reduced flow were detected contralaterally. In the ischemic hemisphere, lCBF showed a marked flow gradient from cingulate to frontolateral cortex, with values ranging from 0.01–0.80 ml/g/min. The epicenter of the most severe flow reduction was in the lateral frontoparietal cortex and was surrounded by zones of more moderate flow reduction. In the ischemic hemisphere, a volume of 117 ± 55 mm³ exhibited lCBF values below 20 % of contralateral control, which corresponded closely to the volume of completed histopathologic infarct previously documented in this model [10]. Flow values in the penumbral (borderzone) range (i.e., lying between 20–40 % of contralateral values) were present in a tissue volume of 136 ± 30 mm³. These data thus yield the surprising insight that *the volume of the initial ischemic penumbra is fully as large as the volume of the severely ischemic core* during the first 1–2 h following MCA occlusion.

lCMR of the ischemic hemisphere showed metabolic suppression below 50 % of the contralateral value in a volume of 85 ± 43 mm³. Within the ischemic penumbra (as defined by the above lCBF criteria), lCMRgl proved to be remarkably preserved at levels of 30–75 µmol/100g/min. Individual lCMRgl brains revealed foci of hypermetabolism in over one-half of animals, with occasional values as high as 160 µmol/100g/min, but with marked inter-animal and inter-regional heterogeneity. Comparative analysis of the relationship between hemispheric brain volumes showing lCBF reduction and metabolic suppression revealed that, for comparable percentages of reduction, significantly larger volumes of the ischemic hemisphere were affected by lCBF reduction than by depressed lCMRgl – signifying marked metabolism-flow uncoupling. This disparity increased markedly at higher flow ranges [1].

The mean CMRgl/CBF ratio in the non-ischemic hemisphere, as determined in the averaged ratio data set, was 51.0 ± 28.7 µmol/100ml [1]. We thus defined values over 108 µmol/100ml (i.e., values exceeding the upper 95 % confidence limit of the mean ratio) to represent significant increases. While there were no foci of increased ratio in the non-ischemic hemisphere, the ischemic hemisphere showed significant metabolism-flow uncoupling. Within the ischemic penumbra (defined as lCBF 20–40 % of control), the mean CMRgl/CBF ratio increased nearly five-fold to 234 ± 100 µmol/100ml and ranged between 108 and 490 µmol/100 ml. These findings, taken together, indi-

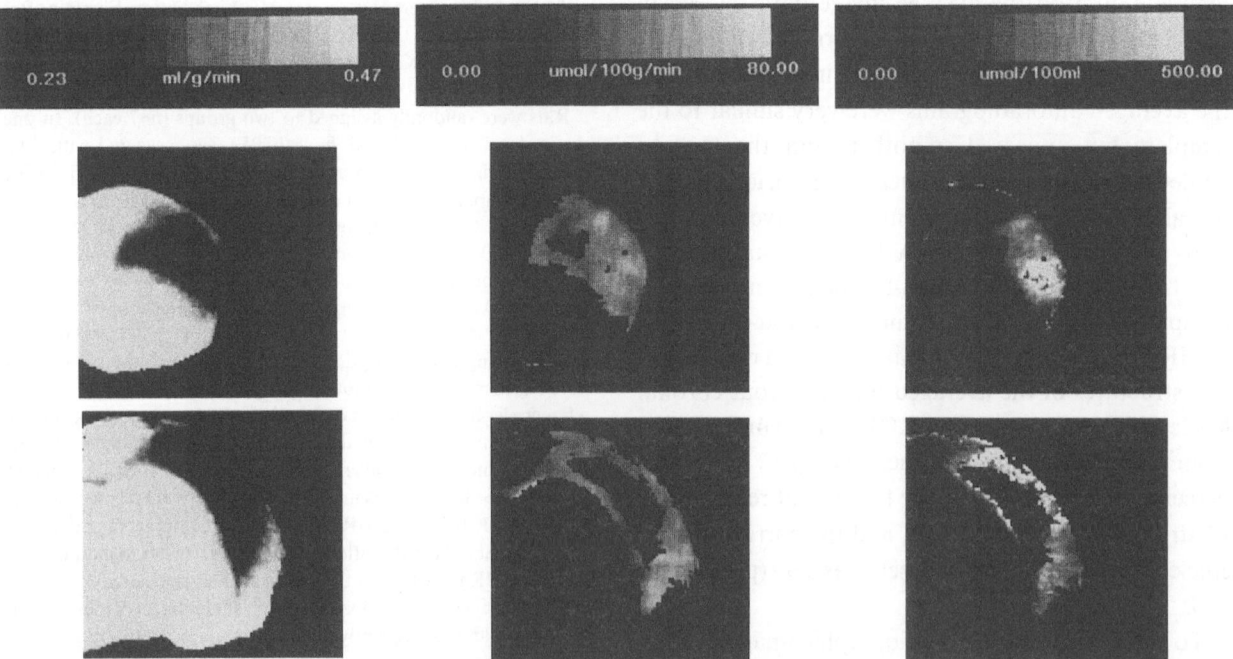

Fig. 2. Thresholded autoradiographic images derived from *averaged* 3D data sets of rats with distal MCA occlusion. Left column: lCBF, thresholded between 20 and 40 % of control. Center column: lCMRgl values displayed for those pixels defined in the left column (lCBF 20–40 % of control). Right column: CMR/CBF ratio image for those pixels defined by the 20–40 % lCBF criterion. Upper row: Anterior pole of the ischemic lesion. Lower row: Center of the ischemic lesion. Penumbral pixels demonstrate well-preserved glucose utilization and markedly elevated CMR/CBF ratio values, averaging nearly five-fold above control values

cate that lCMRgl is well maintained in the early ischemic borderzone despite lCBF values close to the threshold of ischemic injury. At lCBF values above 0.3 ml/g/min, glucose hypermetabolism is observed, accounting for the extreme degree of lCMRgl/CBF ratio elevations. Figure 2 shows representative illustrations.

These results attest to the fact that 3D image-averaging permits powerful insights into the 3D topography of the ischemic penumbra, specifically: 1) that the *volume* of the early penumbra (as defined on the basis of an lCBF range) is substantial, and that the penumbra forms a "shell" around the ischemic core and is particularly concentrated at the anterior and posterior poles of the core; and 2) that, in spite of considerable metabolic heterogeneity within the acute penumbra, this zone nonetheless displays a highly consistent pattern of extremely marked (nearly 5-fold) metabolism/flow uncoupling.

On the basis of correlative electrophysiological studies in this [1] and other models [2,8], it is well established that the penumbra is the site of recurrent ischemic depolarizations which result in marked bioenergetic demands imposed by the need to reestablish ionic equilibria [4].

To summarize, 3D image-averaging techniques offer powerful advantages over conventional approaches in defining the quantitative metabolic and hemodynamic topography of evolving ischemic lesions. Particular advantages include the opportunity to derive *averaged* image data sets based upon replicate studies; to calculate *volumes* of deranged metabolism or blood flow; and finally, to perform arithmetic manipulations (e.g., computation of the CMR/CBF ratio) on averaged 3D imaged data sets. This general approach offers great promise for the study of focal cerebral activation and the effects of pharmacologic agents upon the normal and abnormal brain.

Acknowledgement

This study was supported by Grant NS05820 of the National Institutes of Health.

References

1. Back T, Zhao W, Ginsberg MD (1995) Three-dimensional image-analysis of brain glucose metabolism/blood flow uncoupling and its electrophysiological correlates in the acute ischemic penumbra following middle cerebral artery occlusion. J Cereb Blood Flow Metab 15: 566–577
2. Chen Q, Chopp M, Bodzin G, Chen H (1993) Temperature modulation of cerebral depolarization during focal cerebral ischemia in rats: correlation with ischemic injury. J Cereb Blood Flow Metab 13: 389–394
3. Evans AC, Beil C, Marrett S, Thompson CJ, Hakim A (1988) Anatomical-functional correlation using an adjustable MRI-based region of interest atlas with positron emission tomography. J Cereb Blood Flow Metab 8: 513–530
4. Ginsberg MD, Pulsinelli WA (1994) Editorial: The ischemic penumbra, injury thresholds, and the therapeutic window for acute stroke. Ann Neurol 36: 553–554
5. Henderson TC, Triendl EE, Winter R (1985) Edge- and shape-based geometric registration. IEEE Trans Geosci Remote Sensing GE-23: 334–342
6. Hibbard LS, McGlone JS, Davis DW, Hawkins RA (1987) Three-dimensional representation and analysis of brain energy metabolism. Science 236: 1641–1642
7. Hossmann K-A (1994) Neurological progess: viability thresholds and the penumbra of focal ischemia. Ann Neurol 36: 557–565
8. Iijima T, Mies G, Hossmann K-A (1992) Repeated negative DC deflections in rat cortex following middle cerebral artery occlusion are abolished by MK-801: effect on volume of ischemic injury. J Cereb Blood Flow Metab 12: 727–733
9. Maguire GQ Jr, Noz ME, Lee EM, Schimpf JH (1985) Correlation methods for tomographic images using two or three dimensional techniques. Proc Ninth Internat Conf Information Processing in Medical Imaging, Washington, pp 266–279
10. Markgraf CG, Kraydieh S, Prado R, Watson BD, Dietrich WD, Ginsberg MD (1993) Comparative histopathologic consequences of photothrombotic occlusion of the distal middle cerebral artery in Sprague-Dawley and Wistar rats. Stroke 24: 286–292
11. Paxinos G, Watson C (1982) The rat brain in stereotaxic coordinates. Academic Press, Sydney
12. Sakurada O, Kennedy C, Jehle J, Brown JD, Carbin GL, Sokoloff L (1978) Measurement of local cerebral blood flow with iodo(^{14}C)antipyrine. Am J Physiol 234: H59–H66
13. Sokoloff L, Reivich M, Kennedy C, DesRosiers MH, Patlak CS, Pettigrew KD, Sakurada O, Shinohara M (1977) The (^{14}C)deoxyglucose method for the measurement of local cerebral glucose utilization: theory, procedure, and normal values in the conscious and anaesthetized albino rat. J Neurochem 28: 897–916
14. Yao H, Ginsberg MD, Eveleth DD, LaManna JC, Watson BD, Alonso OF, Loor JY, Busto R (1995) Local cerebral glucose utilization and cytoskeletal proteolysis as indices of evolving focal ischemic injury in core and penumbra. J Cereb Blood Flow Metab: 15: 398–408
15. Young TY, Gunaseekaran S (1988) A regional approach to tracking 3D motion in an image sequence. In: Huang TS (ed) Advances in computer vision and image processing. JAI, USA, pp 63–99
16. Young TY, Wang YL (1984) Analysis of three dimensional rotation and linear shape changes. Pattern Recognit Lett 2: 239–242
17. Zhao W, Ginsberg MD, Smith DW (1995) Three-dimensional quantitative autoradiography by disparity analysis: theory and application to image- averaging of local cerebral glucose utilization. J Cereb Blood Flow Metab 15: 552–565
18. Zhao W, Young TY, Ginsberg MD (1983) Registration and three-dimensional reconstruction of autoradiographic images by the disparity analysis method. IEEE Trans Med Imag 12: 782–791

Correspondence: Myron D. Ginsberg, M.D., Department of Neurology (D4-5), University of Miami School of Medicine, P.O. Box 016960, Miami, FL 33101, U.S.A.

Acta Neurochir (1996) [Suppl] 66: 50–55
© Springer-Verlag 1996

Origins of Glutamate Release in Ischaemia

T. P. Obrenovitch

Department of Neurology Surgery, Institute of Neurology, London, U.K.

Summary

Using microdialysis coupled to on-line detection of glutamate, and recording electrical activity and field potential at the same tissue site, we have shown that the increase in extracellular glutamate under global penumbral conditions is minor. However, in the border of the ischaemic core, recurrent spreading depression is presumably associated with transient vesicular release of glutamate (exocytosis). With ischaemic insults severe enough to provoke anoxic depolarization, such as in the ischaemic core, exocytosis only occurred for a few minutes because it requires ATP hydrolysis, and the magnitude of this release was minor in comparison with that of the total glutamate efflux. Subsequent experiments with a selective inhibitor of high-affinity glutamate transporters suggested that reversal of glutamate uptake may not be a major contributor to the sustained release of glutamate in this condition.

These results, and other considerations, do not favour the view that presynaptic glutamate release and reversed glutamate uptake are suitable targets for neuroprotection in ischaemia. Acting postsynaptically to inhibit recurrent spreading depression (NMDA-receptor antagonists) or to modulate long-lasting enhancement of synaptic efficiency ('*anoxia-induced long-term potentiation*') appear to be more rational strategies.

Keywords: Cerebral ischaemia; glutamate release; excitotoxicity; microdialysis.

Introduction

It is now accepted that excessive opening of glutamate-operated ion channels plays a major role in ischaemia-induced neuronal death, and one possible triggering factor may be increased extracellular glutamate [28,43]. This article summarizes recent developments which have improved our insight into the origins of glutamate release in ischaemia, and discusses their implication for neuroprotection in stroke. It focuses on data obtained by microdialysis with selected models of cerebral ischaemia; for a recent analysis of *in vitro* data, see ref. [32].

Possible Origins of Glutamate Efflux in Ischaemia

Under resting conditions, the level of glutamate in the cytoplasm of brain cells is around 10,000 times higher than that in the extracellular space, a gradient maintained by acidic amino acid carriers present in both presynaptic and glial plasma membranes [29]. As glutamate is the major excitatory transmitter in the brain, these carriers are essential for terminating the postsynaptic action of neurotransmitter glutamate. They are characterized by a high (2–50 μM) affinity for glutamate, and a unique coupling to Na^+ and K^+ [16]. Transport of glutamate into presynaptic terminals or neuroglia requires the simultaneous presence of external Na^+ and internal K^+, and its efficacy is dependent on the Na^+/K^+ gradient across the plasma membrane

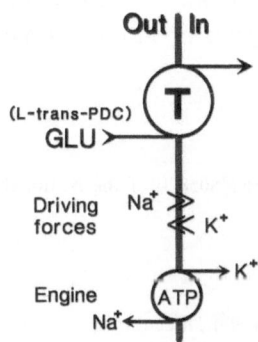

Fig. 1. High-affinity, Na^+-dependent L-glutamic acid uptake. Glutamate influx through this carrier-mediated system requires the simultaneous presence of external Na^+ and internal K^+, and the level of transport is determined by the gradients of Na^+ (out>in) and K^+ (in>out) [17]. This transport mechanism is reversed by increasing extracellular K^+ [17,44]. L-*trans*-PDC (L-*trans*-pyrrolidine-2,4-dicarboxylate) is a selective inhibitor of high-affinity glutamate carriers, but which also releases glutamate by heteroexchange because it is "transportable" [12]

[17] (Fig. 1). The plasmalemnal carriers are different from the transporter which concentrates further glutamate into presynaptic vesicles. Vesicular loading is Na^+-independent, and driven by the internal positive membrane potential that is generated by a vesicular ATPase pumping protons into the vesicle [25].

As with the mechanism for GABA reuptake [39], glutamate uptake can be reversed by high external K^+ [17,44]. In glial cells isolated form the retina of salamander, raising external K^+ to 10 mM was sufficient to reverse glutamate uptake; this effect was activated by intracellular glutamate and Na^+, and increased by membrane depolarization [44]. The reversal of glutamate uptake has two important features: (i) It results from a change in the driving forces (i.e. ionic gradients) (Fig. 1) and not from an alteration of the carrier protein; (ii) the resulting efflux of glutamate is of cytoplasmic origin (metabolic pool of glutamate) and therefore Ca^{2+}-independent. As cerebral ischaemia implies deficiency of the Na^+/K^+ ATPase [8] leading to disruption of the transmembrane ionic gradients with anoxic depolarization [13], moderate insults are likely to produce an imbalance between glutamate efflux and uptake [4], whereas more severe insults may reverse glutamate uptake processes [44].

Another potential source of glutamate efflux in ischaemia is exocytosis of neurotransmitter glutamate, i.e. release from presynaptic vesicles (Fig.1). A number of arguments indicate that neurotransmitter glutamate is released by exocytosis: The Ca^{2+}-dependency of glutamate release from synaptosomes [30], the presence of glutamate in synaptic vesicles [42], and the effective accumulation of glutamate in highly purified synaptic vesicles [25]. In addition, tetanus and botulinum neurotoxins, which selectively cleave proteins mediating the docking of vesicles with the presynaptic plasma membrane [3], markedly inhibited depolarization-induced glutamate release from both synaptosomes [3,27] and cultured cerebellar granule cells [7,46].

All exocytotic processes share a feature which is critical within the context of ischaemia-energy dependency. Vesicular release in response to a stimulus is tightly regulated, mainly at the fusion step, and this step involves ATP hydrolysis [41] (Fig. 2). As ATP rapidly decreases during ischaemia [24,31], exocytosis of glutamate is likely to be limited in this condition.

Cellular swelling subsequent to ischaemia [18,19,22] may also contribute to increasing extracellular glutamate concentration because hypotonic-induced

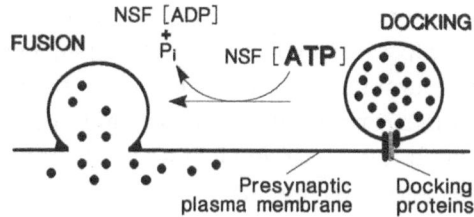

Fig. 2. Exocytosis or vesicular release. Neurotransmitters (glutamate) are stored in synaptic vesicles of which a subpopulation is docked to the 'active zone' at the presynaptic plasma membrane. Stimulation of the nerve terminal leads to an influx of Ca^{2+} (*Ca-dependency*), triggering fusion and neurotransmitter release. The critical step from docking to fusion requires ATP-hydrolysis, and the ATPase activity may reside in the N-ethylmaleimide-sensitive fusion (NSF) protein itself [41]. P_i, inorganic phosphate

swelling of cultured astrocytes caused release of intracellular glutamate [20]. The swelling-induced glutamate release, which did not appear to involve reversal of the Na^+-dependent uptake system, was inhibited by a number of anion transport inhibitors [20], including L-644,711 which improved outcome in an experimental brain trauma/hypoxia model [21]. Finally, in addition to the potential mechanisms outlined above, cellular lysis may also be responsible for glutamate leakage and high extracellular glutamate levels in ischaemia.

Materials and Methods

Our methods were described in detail in original publications [36,37,47]. Briefly, changes in extracellular glutamate concentrations were measured using intracerebral microdialysis [2] in anaesthetized rats. Dialysate glutamate levels were measured by HPLC [36], or by on-line enzymatic analysis whenever optimal time resolution was required [34,49]. In all our experiments, EEG and extracellular direct current (DC) potential were recorded with a chlorided silver wire incorporated within the microdialysis probe [33]. This allowed us to correlate changes in extracellular glutamate with those in electrical activity and ionic homeostasis, i.e. with the severity of ischaemia.

Selected animal models were used depending on the hypotheses to be tested: (i) Graded forebrain ischaemia produced by partial 4-vessel occlusion [36]; (ii) Cortical spreading depression (SD) evoked by microdialysis application of high K^+ [37], to investigate whether glutamate may be released during recurrent SD which propagates from the ischaemic core to surrounding regions [14]; and (iii) proximal middle cerebral artery occlusion (MCAO) or cardiac arrest to examine the time course of changes in extracellular glutamate with severe ischaemia rapidly producing anoxic depolarization [47].

Glutamate Release in the Ischaemic Penumbra

Determination of glutamate efflux in the penumbra (i.e. regions where electrical activity is abolished but ionic homeostasis preserved) [1] is important, because glutamate receptor antagonists reduced the volume of infarction in experimental focal ischaemia [26] but failed to be neuroprotective in models of transient severe global ischaemia [5,10]. As the penumbra is a narrow region in the rat MCAO [45] whose distribution may be perturbed by the implantation of a microdialysis probe [2], we have modified the 4-vessel occlusion rat model of cerebral ischaemia to produce sustained and controlled periods of penumbral ischaemia in the forebrain. During 30 min of penumbral ischaemia, extracellular concentrations of excitatory amino acids in the striatum increased slightly but did not reach critical levels. Massive overflow of neuroactive compounds only occurred with sustained anoxic depolarization [36] (Fig. 3). These data strongly suggest that excitotoxic processes in the penumbra are not related to changes in the extracellular level of excitatory amino acids generated *within* this region. With focal ischaemia however, excessive glutamate could leak from the ischaemic core, damage nearby neurons with further efflux of glutamate, extending tissue damage according to a self-propagating event.

Careful examination of glutamate toxicity in rat cerebellar slices has invalidated this hypothesis; as long as the transmembrane Na^+ and K^+ gradients were maintained, which is the case in the penumbra, high glutamate in the incubation medium was only toxic to the outermost regions of the slice [11].

An alternative cause of glutamate efflux in the penumbra is recurrent SD, which propagates in regions adjacent to the ischaemic core and contributes to the extent of tissue damage [14]. Glutamate release has long been suspected to occur during SD because this event is associated with Ca^{2+}-influx, and both SD triggering and propagation require functional NMDA-receptor ionophore complexes [23]. Previous attempts to demonstrate glutamate efflux with SD did not provide conclusive evidence [9,40] because microdialysis inhibited SD propagation by buffering the transient increase in extracellular K^+ associated with it [38]. We showed that increasing the K^+ concentration in the perfused artifical cerebrospinal fluid (CSF) dose-dependently restored SD propagation and revealed a large, synchronous transient increase in extracellular glutamate [37,38] (Fig. 4). As both SD elicitation and propagation are Ca^{2+}-dependent [15], glutamate released during SD is probably of vesicular origin.

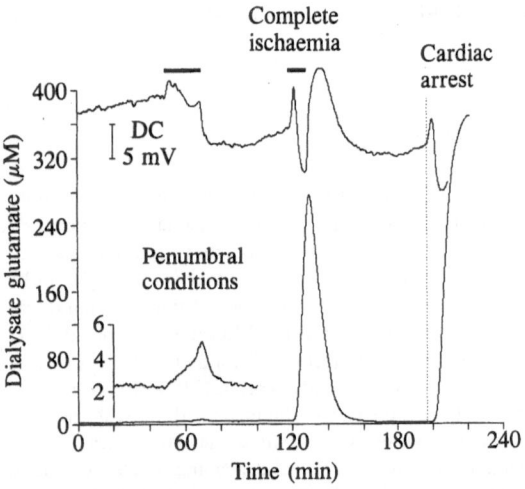

Fig. 3. Changes in dialysate glutamate during cerebral ischaemia of graded severity (bottom traces). Microdialysis probes, incorporating an electrode for the recording of EEG and DC potential (upper trace), were implanted in the striatum of rats anaesthetized with halothane. Dialysate glutamate concentration was continuously monitored by enzyme-fluorescence [34]. Two to three hours after implantation, animals were subjected to the following procedure: (i) 20 min of penumbral conditions (i.e. EEG silence without anoxic depolarization) produced by controlled 4-vessel occlusion [36]; (ii) 40 min of recirculation; (iii) 5 min of complete ischaemia; (iv) >60 min recirculation before cardiac arrest. Note that marked glutamate efflux only occurred during complete ischaemia with anoxic depolarization. Data are from a single representative experiment. The bottom left insert shows the changes in dialysate glutamate during penumbral conditions; its scale is 20 fold expanded in comparison with that used for representing glutamate changes throughout the procedure

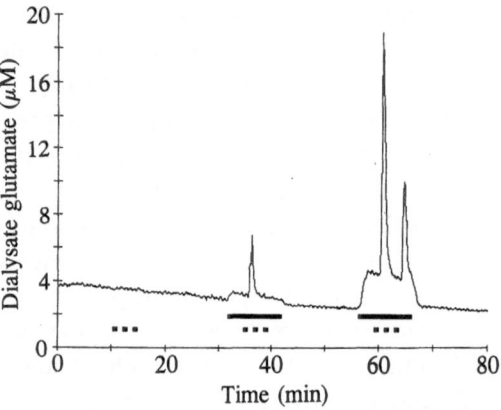

Fig. 4. Representative glutamate release evoked by propagating SD. Microdialysis probes were implanted in the cerebral cortex of rats anaesthetized with halothane, and dialysate glutamate was monitored by enzyme-amperometry for optimal time resolution [49]. SD was elicited by application of K^+ 3 mm in front of the recording probe (horizontal dotted lines). SD did not propagate through the regions immediately surrounding the microdialysis probe when it was perfused with control artificial CSF (left K^+-stimulus). Increasing the concentration of K^+ in the perfusion medium to 30 and 60 mM (middle and right stimulus, respectively; horizontal solid lines) dose-dependently restored SD propagation and showed marked glutamate release synchronous with SD. The small sustained increases in dialysate glutamate were a direct effect of 30 and 60 mM K^+ application. Data are from Obrenovitch and coworkers [38], with permission of the authors and Raven Press Ltd

Time Course and Origins of Glutamate Release in the Ischaemic Core

The time course of changes in extracellular glutamate and their Ca^{2+}-dependency were studied in the rat striatum during focal cerebral ischaemia using microdialysis coupled to flow analysis of the dialysate [47]. When the probe was perfused with control artificial CSF, ischaemia produced a biphasic increase in extracellular glutamate which started from the onset of ischaemia. During the first phase lasting around 10 min, dialysate glutamate increased by around 30 µM. Then, dialysate glutamate increased progressively to its maximum (80–85 µM), reached after 55–60 min of ischaemia, where it remained for at least 3 h [47]. A similar pattern of changes was obtained during complete ischaemia produced by cardiac arrest, with a probe implanted in the rat cerebral cortex and dialysate glutamate recorded by enzyme-amperometry (Fig. 5).

The early component of the extracellular glutamate kinetic was no longer detectable when Ca^{2+} was omitted from the perfusing medium [47], suggesting that it originated from vesicular glutamate. This implies that residual ATP was maintained at a level sufficient to sustain exocytotic release for several minutes after ischaemia onset, probably at the expense of phospho-creatine whose depletion precedes that of ATP [31]. Experiments with cardiac arrest (Fig. 5) showed that exocytotic release during ischaemia was closely associ-ated with anoxic depolarization, an event which combines massive Ca^{2+} influx [13] with residual ATP levels of around 1/3 normal [35].

In order to test whether the second phase of glutamate release in ischaemia was due to reversal of its uptake, the effect of selective blockade of high-affinity glutamate transporters was studied, using L-*trans*-pyrrolidine-2,4-dicarboxylate (L-*trans*-PDC), a new selective inhibitor of these transporters [12]. L-*trans*-PDC was applied through the microdialysis probe, starting 10 min before cardiac arrest and continued throughout the post-mortem recording period. Although 2.5 mM L-*trans*-PDC markedly increased basal levels of dialysate glutamate (Fig. 6) the time course of postmortem glutamate changes was barely altered. The maximum rate of release during the first exocytotic phase, and the increase in dialysate glutamate, appeared slightly exacerbated (Table 1), but the kinetic of the second, Ca^{2+}-independent phase was essentially unaltered (Fig. 6). The latter observation does not support that reversal of glutamate uptake is a major contributor to increased extracellular glutamate in ischaemia, but this finding must be inter-preted with caution, because L-*trans*-PDC is a com-petitive (i.e. transportable) inhibitor of the glutamate transporters which also evokes glutamate release by heteroexchange [12,48].

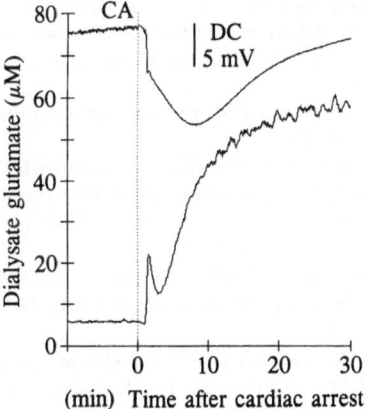

Fig. 5. Biphasic changes in dialysate glutamate (bottom trace) and corresponding DC potential (upper trace) during complete ischaemia produced by cardiac arrest (CA). Microdialysis probes were implanted in the cortex of rats anaesthetized with halothane, and dialysate glutamate recorded by enzyme-amperometry. In this experiment, the first (exocytotic) phase of glutamate release was especially marked; note that it was closely associated with anoxic depolarization. Data are from Zilkha and coworkers [49], with per-mission of the authors and Elsevier Publishers

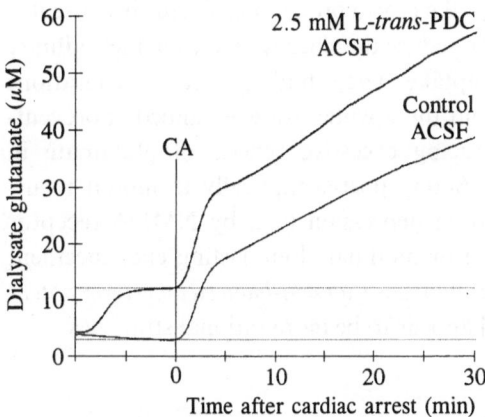

Fig. 6. Glutamate release produced by cardiac arrest (CA): Effect of blockade of high-affinity glutamate transporters. Microdialysis probes were implanted in the cortex of rats anaesthetized with halothane, and dialysate glutamate recorded by enzyme-fluorescence [34]. In 6 experiments, 2.5 mM of L-*trans*-PDC (a selective inhibitor of high-affinity glutamate transporters) were added to the perfusion medium, starting 10 min before cardiac arrest (upper trace). In the control group (n = 6), control artificial CSF was perfused throughout the procedure (lower trace). Although L-*trans*-PDC consistently produced a marked step increase in the basal levels of dialysate glutamate, it did not markedly alter the kinetic of glutamate release produced by cardiac arrest. Each trace is the average of 6 experiments

Table 1. *Effect of Glutamate Uptake Inhibition (2.5 mM L-Trans-PDC in ACSF) on the First (Exocytotic) Phase of Glutamate Release Following Cardiac Arrest*

	Time after cardiac arrest for maximum rate of release (sec)	Maximum rate of release (μM min^{-1})	Change in dialysate level at the end of this phase (μM)
Control ACSF	157 ± 8	5.5 ± 0.6	14.2 ± 0.8
L-*trans*-PDC	150 ± 6	7.1 ± 0.5	17.8 ± 1.4
ACSF	P = 0.53	P = 0.055	P = 0.053

Values are mean ± S.E.M., n = 6; *P* comparison to the control group by Student's test.
ACSF artificial cerebrospinal fluid.

Conclusion and Comments

Our findings strongly suggest that the ischaemic penumbra is not exposed to sustained increased extracellular levels of glutamate. Marked glutamate release, presumably of vesicular origin, only occurs transiently with propagating SD. During severe ischaemia with anoxic depolarization, the increase of glutamate levels in the extracellular space is biphasic, and most of the released glutamate is of non-vesicular origin, probably from both neurons and glia. In addition, selective inhibition of presynaptic vesicular glutamate release may be difficult because the synaptic machinery for release of neurotransmitter glutamate is not unique to glutamatergic synapses [27], nor to synaptic transmission in general, but common to a wide range of secretory and membrane-fusion processes [41]. Interfering with glutamate transporters to reduce release through reversed uptake is also prone to problems because it is likely to also reduce the efficacy of normal high-affinity glutamate uptake. These findings and considerations do not favour therapeutic strategies aimed at preventing or reducing excessive release of glutamate in ischaemia. Acting post-synaptically to inhibit recurrent spreading depression (e.g. by NMDA-receptor antagonists) or modulate long-lasting enchancement of synaptic efficiency ('*anoxia-induced long-term potentiation*') [6] appear to be more rational strategies.

References

1. Astrup J, Siesjö BK, Symon L (1981) Threshold in cerebral ischemia- the ischemic penumbra. Stroke 12: 723–725
2. Benveniste H, Hüttemeier PC (1990) Microdialysis–theory and application. Prog Neurobiol 35: 195–215
3. Blasi J, Chapman ER, Link E, Binz T, Yamasaki S, De Camilli P, Südhof TC, Niemann H, Jahn R (1993) Botulinum neurotoxin A selectively cleaves the synaptic protein SNAP-25. Nature 365: 160–163
4. Bradford HF, Young AMJ, Crowder JM (1987) Continuous leakage from brain cells is balanced by compensatory high-affinity reuptake transport. Neurosci Lett 81: 296–302
5. Buchan A, Li H, Pulsinelli WA (1991) The N-methyl-D-aspartate antagonist, MK-801, fails to protect against neuronal damage caused by transient, severe forebrain ischemia in adult rats. J Neurosci 11: 1049–1056
6. Crépel V, Hammond C, Krnjevic K, Chinestra P, Ben-Ari Y (1993) Anoxia-induced LTP of isolated NMDA receptor-mediated synaptic responses. J Neurophysiol 69: 1774–1778
7. Didier M, Héaulme M, Gonalons N, Soubrié P, Bockaert J, Pin JP (1993) 35 mM K$^+$-stimulated ^{45}Ca^{2+} uptake in cerebellar granule cell cultures mainly results from NMDA receptor activation. Eur J Pharmacol 244: 57–65
8. Erecinska M, Silver IA (1994) Ions and energy in mammalian brain. Prog Neurobiol 43: 37–71
9. Fabricius M, Hensen LH, Lauritzen M (1993) Microdialysis of interstitial amino acids during spreading depression and anoxic depolarization in rat neocortex. Brain Res 612: 61–69
10. Fleischer JE, Tateishi A, Drummond JC, Scheller MS, Grafe MR, Zornow MH, Shearman GT, Shapiro HM (1989) MK-801, an excitatory amino acid antagonist, does not improve neurologic outcome following cardiac arrest in cats. J Cereb Blood Flow Metab 9: 795–804
11. Garthwaite G, Williams GD, Garthwaite J (1992) Glutamate toxicity: an experimental and theoretical analysis. Eur J Neurosci 4: 353–360
12. Griffiths R, Dunlop J, Gorman A, Senior J, Grieve A (1994) L-*trans*-pyrrolidine-2, 4-dicarboxylate and *cis*-1-aminocyclobutane-1,3-dicarboxylate behave as transportable, competitive inhibitors of the high-affinity glutamate transporters. Biochem Pharmacol 47: 267–274
13. Hansen AJ (1985) Effects of anoxia on ion distribution in the brain. Physiol Rev 65: 101–148
14. Iijima T, Mies G, Hossmann K-A (1992) Repeated negative DC deflections in rat cortex following middle cerebral artery occlusion are abolished by MK-801: effect on volume of ischemic injury. J Cereb Blood Flow Metab 12: 727–733
15. Jing J, Aitken PG, Somjen GG (1993) Role of calcium channels in spreading depression in rat hippocampal slices. Brain Res 604: 251–259
16. Kanai Y, Smith CP, Hediger MA (1993) The elusive transporters with a high-affinity for glutamate. Trends Neurosci 16: 365–370
17. Kanner BI, Bendahan A (1982) Binding order of substrates to the sodium and potassium ion coupled L-glutamic acid transporter from rat brain. Biochem 21: 6327–6330
18. Kempski O, Staub F, Jansen M, Schödel F, Baethmann A (1988) Glial swelling during extracellular acidosis in vitro. Stroke 19: 385–392
19. Kempski O, Zimmer M, Neu A, v Rosen F, Jansen M, Baethmann A (1987) Control of glial cell volume in anoxia: In vitro studies on ischemic cell swelling. Stroke 18: 623–628

20. Kimelberg HK, Goderie SK, Higman S, Pang S, Waniewski RA (1990) Swelling-induced release of glutamate, aspartate, and taurine from astrocyte cultures. J Neurosci 10: 1583–1591

21. Kimelberg HK, Rose JW, Barron KD, Waniewski RA, Cragoe EJ (1989) Astrocytic swelling in traumatic-hypoxic brain injury. Beneficial effects of an inhibitor of anion exchange transport and glutamate uptake in glial cells. Molecul Chem Neuropathol 11: 1–31

22. Korf J, Postema F (1988) Rapid shrinkage of rat striatal extracellular space after local kainate application and ischemia as recorded by impedance. J Neurosci Res 19: 504–510

23. Lauritzen M (1994) Pathophysiology of the migraine aura: the spreading depression theory. Brain 117: 199–210

24. Lowry OH, Passonneau JV, Hasselberger FX, Schultz D (1964) Effects of ischemia on known substrates and cofactors of the glycolytic pathway in brain. J Biol Chem 239: 18–30

25. Maycox PR, Hell JW, Jahn R (1990) Amino acid neurotransmission: spotlight on synaptic vesicles. Trends Neurosci 13: 83–87

26. McCulloch J, Ozyurt E, Park CK, Nehls DG, Teasdale GM, Graham DI (1993) Glutamate receptor antagonists in experimental focal cerebral ischaemia. In: Baethmann A, Kempski O, Schürer L (eds) Mechanisms of secondary brain damage -current state. Acta Neurochir Wien [Suppl] 57: 73–79

27. McMahon HT, Foran P, Dolly JO, Verhage M, Wiegant VM, Nicholls DG (1992) Tetanus toxin and botulinum toxins A and B inhibit glutamate, gamma-aminobutyric acid, aspartate, and metenkephalin release from synaptosomes. Clues to the locus of action. J Biol Chem 267: 21338–21343

28. Meldrum BS, Millan MH, Obrenovitch TP (1993) Injury induced excitatory amino acid release. In: Globus MYT, Dietrich WD (eds) The role of neurotransmitters in brain injury. Plenum, New York, pp 1–7

29. Nicholls DG, Attwell D (1990) The release and uptake of excitatory amino acids. Trends Neurosci 11: 462–468

30. Nicholls DG, Sihra T (1986) Synaptosomes possess an exocytotic pool of glutamate. Nature 321: 772–773

31. Obrenovitch TP, Garofalo O, Harris RJ, Bordi L, Ono M, Momma F, Bachelard HS, Symon L (1988) Brain tissue concentration of ATP, phosphocreatine, lactate, and tissue pH in relation to reduced cerebral blood flow following experimental acute middle cerebral artery occlusion. J Cereb Blood Flow Metab 8: 866–874

32. Obrenovitch TP, Richards DA (1995) Neurotransmitter changes in the extracellular fluid in cerebral ischaemia. Cerebrovasc Brain Metabol Rev 7: 1–54

33. Obrenovitch TP, Richards DA, Sarna GS, Symon L (1993a) Combined intracerebral microdialysis and electrophysiological recording: methodology and applications. J Neurosci Meth 47: 139–145

34. Obrenovitch TP, Sarna, GS, Millan MH, Lok S-Y, Kawauchi M, Ueda Y, Symon L (1990a) Intracerebral dialysis with on-line enzyme fluorometric detection: a novel method to investigate the changes in the extracellular concentration of glutamic acid. In: Krieglstein J, Oberpichler H (eds) Pharmacology of cerebral ischemia 1990. Thieme, Stuttgart, pp 23–31

35. Obrenovitch TP, Scheller D, Matsumoto T, Tegtmeier F, Höller M, Symon L (1990b) A rapid redistribution of hydrogen ions is associated with depolarization and repolarization sub-

sequent to cerebral ischaemia/reperfusion. J Neurophysiol 64: 1125–1133

36. Obrenovitch TP, Urenjak J, Richards DA, Ueda Y, Curzon G, Symon L (1993b) Extracellular neuroactive amino acids in the rat brain striatum during moderate and severe transient ischemia. J Neurochem 61: 178–186

37. Obrenovitch TP, Zilkha E (1995) Changes in extracellular glutamate concentration associated with propagating cortical spreading depression. In: Olesen J, Moskowitz MA (eds) Experimental headache models in animal and man. Raven, New York, pp 113–117

38. Obrenovitch TP, Zilkha E, Urenjak J (1995) Intracerebral microdialysis: electrophysiological evidence of a critical pitfall. J Neurochem 64: 1884–1887

39. Pin J-P, Bockaert J (1989) Two distinct mechanisms differentially affected by excitatory amino acids trigger gamma-aminobutyric acid release from striatal neurones in primary cultures. J Neurosci 9: 648–656

40. Scheller D, Heister U, Kolb J, Tegtmeier F (1993) On the role of excitatory amino acids during generation and propagation of spreading depression. In: Lemenkühler A, Grotemeyer KH, Tegtmeier F (eds) Migraine: basic mechanisms and treatments. Urban and Schwarzenberg, München, pp 355–356

41. Söllner T, Rothman JE (1994) Neurotransmission: harnessing fusion machinery at the synapse. Trends Neurosci 17: 344–349

42. Storm-Mathisen J, Leknes AK, Bore AT, Waaland JL, Edminson P, Haug F-MS, Ottersen OP (1983) First visualization of glutamate and GABA in neurones by immunocytochemistry. Nature 301: 517–520

43. Szatkowski M, Attwell D (1994) Triggering and execution of neuronal death in brain ischaemia: two phases of glutamate release by different mechanisms. Trends Neurosci 17: 359–365

44. Szatkowski M, Barbour B, Attwell D (1990) Non-vesicular release of glutamate from glial cells by reversed electrogenic glutamate uptake. Nature 348: 443–446

45. Tyson GW, Teasdale GM, Graham DI, McCulloch J (1984) Focal cerebral ischemia in the rat: topography of hemodynamic and histopathological changes. Ann Neurol 15: 559–567

46. Van-Vliet BJ, Sebben M, Dumuis A, Gabrion J, Bockaert J, Pin JP (1989) Endogenous amino acid release from cultured cerebellar neuronal cells: effect of tetanus toxin on glutamate release. J Neurochem 52: 1229–1239

47. Wahl F, Obrenovitch TP, Hardy AM, Plotkine M, Boulu R, Symon L (1994) Extracellular glutamate during focal cerebral ischaemia in rats: Time course and calcium-dependency. J Neurochem 63: 1003–1011

48. Waldmeier PC, Wicki P, Feldtrauer J-J (1993) Release of endogenous glutamate from rat cortical slices in presence of the glutamate uptake inhibitor L-trans-pyrrolidine-2,4-dicarboxylic acid. Naunyn Schmiedeberg's Arch Pharmacol 348: 478–485

49. Zilkha E, Obrenovitch TP, Koshi A, Kusakabe H, Bennetto HP (1995) Extracellular glutamate: on-line monitoring using microdialysis coupled to enzyme-amperometric analysis. J Neurosci Meth 60: 1–9

Correspondence: T. P. Obrenovitch, Ph.D., Gough-Cooper Department of Neurological Surgery, Institute of Neurology, Queen Square, London, WC1N 3BG, U.K.

Acta Neurochir (1996) [Suppl] 66: 56–62
© Springer-Verlag 1996

Swelling, Intracellular Acidosis, and Damage of Glial Cells

F. Staub[1], **A. Winkler**[1], **J. Haberstok**[1], **N. Plesnila**[1], **J. Peters**[1], **R. C. C. Chang**[1], **O. Kempski**[2], and **A. Baethmann**[1]

[1]Institut für Chirurgische Forschung, Klinikum Großhadern der Ludwig-Maximilians-Universität München, München, and [2]Institut für Neurochirurgische Pathophysiologie, Klinikum der Johannes-Gutenberg Universität Mainz, Mainz, Federal Republic of Germany

Summary

Cerebral ischemia and severe head injury among others are associated with a limited availability of oxygen, leading to cell catabolism as well as anaerobic glycolysis. Resulting metabolites, such as arachidonic- and lactic acid, can be expected to leak into perifocal brain areas, contributing there to cytotoxic swelling and damage of neurons and glia. Since elucidation of mechanisms underlying cell swelling and damage in the brain is difficult in vivo, respective investigations were carried out in vitro using suspended glial cells. Thereby, effects of arachidonic acid (AA) and of lactacidosis on glial cell volume, intracellular pH (pH_i), and cell damage were analyzed utilizing flow cytometry. AA led to an immediate, dose dependent swelling and intracellular acidosis of glial cells. A concentration of 0.1 mM increased cell volume to 110 % of control and decreased pH_i to 7.05. Whereas glial swelling was permanent, pH_i recovered to baseline after 90 min. Cell viability of 90 % remained unchanged after addition of AA up to 0.1 mM, while at 0.5 mM it was significantly decreasing. Glial swelling from AA was nearly completely inhibited by the aminosteroid U-74389F or by using a Na^+-free suspension medium for the experiment. Acidification of the medium to pH 6.8 or 6.2 led to a cell volume of 110 % or 120 % of control without affecting cell viability. The cells were not capable to defend their normal pH_i during lactacidosis of the suspension medium but became acidotic as well. Addition of amiloride or utilization of Na^+-free medium inhibited cell swelling from lactacidosis, while intracellular acidosis was even more pronounced. The results indicate that AA as well as acidosis are potent mediators of glial swelling and damage at levels found under pathophysiological conditions in the brain in vivo. Whereas intracellular acidification caused by AA was reversible, glial cells were unable to regulate their pH_i during maintenance of extracellular acidosis. Concerning the mechanisms of glial swelling by AA, the production of oxygen- and lipid radicals might play a major role in the swelling process. The results indicate a role of the Na^+/H^+-antiporter in acidosis-induced glial swelling, whereas the exchanger has a limited significance for maintenance of pH_i. As seen, the final pathway of glial swelling from both, AA and lactacidosis, requires a net influx of Na^+-ions, probably together with Cl^--ions, and osmotically obliged water.

Keywords: Glial cells; cytotoxic brain edema; arachidonic acid; acidosis.

Introduction

During pathophysiological events in the brain, such as ischemia or trauma, control of glial and neuronal volume is lost. The swelling of cellular elements in the brain – of glial endfeet or dendrites – is referred to as *cytotoxic brain edema* [3]. Development of brain edema and the resulting rise in intracranial pressure are of predominant significance and often determine the ultimate outcome of neurosurgical patients [1]. The mechanisms underlying swelling of brain cells are not completely understood, which may be attributable to the fact that during ischemia or brain trauma a multitude of processes becomes simultaneously activated in a particularly complex tissue. Among others, release and accumulation of free fatty acids as well as development of tissue acidosis are regularly observed during pathophysiological states in the brain [2,4,5,16,23].

The release of free fatty acids involves activation of phospholipases and breakdown of membrane phospholipids. Especially arachidonic acid (AA) as polyunsaturated compound and precursor of prostaglandins, leukotrienes and oxygen-derived free radicals is considered to mediate pathological processes [41]. In cerebral ischemia concentrations of free AA of up to 0.5 mMol/kg may accumulate [23]. Brain injury and seizures are also causing increased levels of free AA [4,5,32]. In addition to fatty acid liberation, pathophysiological events in the brain are associated with an enhancement of anaerobic metabolism and, thereby, development of intra-and extracellular lactacidosis. In cerebral ischemia the extracellular pH may fall below 6.0 with accumulation of lactic acid to a level of 20–30 mMol/kg [16,24]. Because in vitro a higher level of control of experimental conditions is possible, the significance of isolated pathophysiological factors, such as AA or lactacidosis, was analyzed as to the induction of cell swelling, disturbance of pH_i-regulation, and cell damage. Flow cytometrical studies were performed for that purpose using C6 glioma cells or astrocytes from primary culture.

Materials and Methods

C6 glioma cells were cultivated as monolayers in Petri dishes using Dulbecco's modified minimum essential medium (DMEM) buffered with 25 mM bicarbonate. The medium was supplemented with 10 % fetal calf serum (FCS) and 100 IU/ml penicillin G and 50 μg/ml streptomycin. The cells were grown in a humidified atmosphere of 5 % CO_2 and room air at 37 °C. Glial cells from primary culture were prepared from the brain of 3-day-old rats according to a modified method of Frangakis and Kimelberg [9,19]. The culture conditions were identical with those described above. For the experiments only confluent cultures were used. The cells were harvested with 0.05 % trypsin-0.02 % EDTA in phosphate-buffered saline and washed twice thereafter. After resuspension in serum-free medium the glial cells were transferred to a plexiglas incubation chamber supplied with electrodes for control of pH, temperature, and pO_2. A gas-permeable silicon rubber tube in the chamber provided the cell suspension with a mixture of O_2, CO_2, and N_2. Details of the model have been published elsewere [18,19]

The volume of the glial cells was determined by flow cytometry using an advanced coulter system with hydrodynamic focusing [14]. The technique was also employed for measurement of cell viability by exclusion of the fluorescence dye propidium iodide (final concentration: 40 μg/ml) [31,35]. Excitation of the fluorochrome was induced by a Hg-light source following passage through a 500 nm short pass filter. Maximum emission of propidium iodide fluorescence at 630 nm was measured by using a 580 nm long pass filter [35]. A window integration system was employed for discrimination of propidium iodide-positive (dead) from non-fluoresceing (viable) cells [15]. The fluorescence indicator BCECF was utilized for measurement of the cytosolic pH by flow cytometry [27,36]. C6 glioma cells were loaded with BCECF-AM (10 μM) for 20 min at 37 °C in the incubation chamber prior to the experiment. BCECF was excited as described above, fluorescence emission of the cells was measured by employment of 532 and 630 nm band pass filters of 20 nm band width each. The ratio of BCECF fluorescence intensities at these wavelength was utilized as measure of pH_i [36]. pH_i-calibration was made after adjustment of the intra- to extracellular pH by the H^+-ionophore nigericin at a K^+-concentration of 120 mM in the medium [38].

The experiments were performed after a 45-minute control period for measurements of normal cell volume, viability and medium osmolality. Subsequently, the C6 glioma cell suspension was added with AA in a dose range of 0.01 mM to 1.0 mM (final concentrations). Cell volume and viability were examined for 90 min during incubation with AA. In further experiments swelling of C6 glioma cells by AA (0.1 mM) was studied during blocking of lipid peroxidation by the aminosteroid U-74389F (0.1 mM; Upjohn, Kalamazoo, MI, USA) as well as in a Na^+-free medium, where Na^+-ions were replaced by choline and bicarbonate by 10 mM HEPES. pH_i of C6 glioma cells was measured at AA concentrations in the medium of 0.05 or 0.1 mM. In experiments with astrocytes from primary culture, cell volume and viability were studied at 0.05 or 0.1 mM AA.

In another experimental series the role of lactacidosis regarding changes in pH_i and cell volume of glial cells was investigated. Following a control period, the pH in the suspension of astrocytes from primary culture was decreased from 7.4 to 6.8 or 6.2 by addition of isotonic lactic acid (~350 mM). Cell volume, viability, and pH_i were measured for 60 min during lactacidosis and for another 30 min after restoration of the medium pH to 7.4 by addition of isotonic NaOH. The involvement of the Na^+/H^+-antiporter in glial swelling and pH_i-regulation was analyzed by employment of amiloride (0.1 mM) or suspension of the cells in Na^+-free medium, the function of H^+-channels by administration of $ZnCl_2$ (0.1 mM).

Results

Arachidonic Acid

Administration of AA to the suspension led to immediate dose-dependent swelling of C6 glioma cells. Concentrations as low as 0.01 mM were found to increase cell volume to 103.0 ± 1.0 % (mean ± SEM) of control within 10 min ($p < 0.01$, Table 1). A volume increase to 112.2 ± 1.9 % or 121.1 ± 3.7 % of control was obtained, when the cells were administered with 0.1 or 1.0 mM AA, respectively ($p < 0.001$). After a rapid swelling phase the increased cell volume was stable for the remaining observation period (Table 1). Additional experiments were performed to analyze mechanisms of the AA-induced glial swelling. Administration of the aminosteroid U-74389F to scavenge lipid- and lipid peroxide radicals prevented cell swelling practically completely ($p < 0.01$). Similar results were obained when the experiments were conducted with choline for replacement of Na^+-ions in the suspension medium ($p < 0.01$) [36].

Viability of the C6 glioma cells was 90.6 ± 0.8 % under control conditions, and it remained unchanged after addition of AA up to a level of 0.1 mM. At 0.5 mM, the number of viable cells after a delay of 10 min started to significantly decrease, eventually falling to 72.8 ± 1.9 % at 90 min ($p < 0.01$). Upon incubation with AA at 1.0 mM, cell viability began immediately to decline. Only 32.7 ± 4.7 % of the C6 glioma cells were viable at 90 min ($p < 0.001$).

Swelling and damage of C6 glioma cells by AA was confirmed in experiments using astrocytes from primay culture. Addition of AA at a dose level of 0.05 or 0.1 mM led to significant swelling of astrocytes ($p < 0.01$), which was comparable to that of C6 glioma cells. Astrocytes were, however, more vulnerable. AA concentrations as low as 0.1 mM led already to a decrease in cell viability ($p < 0.05$) [36]. The intracellu-

Table 1. *Volume (Top) and pH$_i$ (Bottom) of C6 Glioma Cells During Incubation with Arachidonic Acid*

Time [min]	Control			Incubation with arachidonic acid					
	−15	−10	−5	1	5	10	30	60	90
Control	99.85 ±0.52	99.93 ±0.19	100.22 ±0.41	100.28 ±0.55	100.68 ±0.58	100.36 ±0.49	99.75 ±0.39	100.03 ±0.76	99.78 ±0.87
AA 0.01 mM	100.11 ±0.14	99.93 ±0.10	99.95 ±0.09	102.36 ±1.06	102.76 ±1.47	103.04** ±1.02	104.85** ±1.59	104.12** ±1.11	104.30** ±1.81
AA 0.05 mM	100.00 ±0.14	100.22 ±0.10	99.77 ±0.08	104.15** ±0.82	105.65** ±1.28	105.67** ±1.50	104.41** ±1.38	103.83** ±1.32	105.25** ±1.34
AA 0.1 mM	100.40 ±0.26	99.93 ±0.17	99.67 ±0.25	104.45** ±1.16	110.04** ±1.49	112.17** ±1.90	111.00** ±2.23	109.58** ±2.20	108.85** ±1.72
AA 0.5 mM	99.48 ±0.35	100.73 ±0.53	99.79 ±0.48	106.55** ±2.68	114.92** ±2.99	115.08** ±1.77	115.97** ±3.86	116.22** ±3.59	116.92** ±3.38
AA 1.0 mM	99.73 ±0.28	100.79 ±0.31	99.48 ±0.26	108.07** ±1.32	118.77** ±1.47	121.07** ±3.72	119.94** ±3.23	127.42** ±3.32	123.08** ±6.42
Time [min]	Control			Incubation with arachidonic acid					
	−15	−10	−5	1	5	10	30	60	90
Control	7.27 ±0.02	7.28 ±0.01	7.30 ±0.02	7.31 ±0.02	7.31 ±0.02	7.31 ±0.02	7.32 ±0.01	7.30 ±0.02	7.30 ±0.01
AA 0.05 mM	7.27 ±0.01	7.26 ±0.01	7.26 ±0.01	7.22* ±0.01	7.12* ±0.03	7.13* ±0.03	7.19* ±0.06	7.29 ±0.01	7.31 ±0.05
AA 0.01 mM	7.26 ±0.02	7.26 ±0.02	7.25 ±0.02	7.12** ±0.02	7.06** ±0.03	7.09** ±0.02	7.15** ±0.03	7.24 ±0.02	7.25 ±0.01

* p < 0.05 vs. Control, ** p < 0.01 vs. Control.
Volume response (top) and intracellular pH (bottom) of C6 glioma cells during exposure to different concentrations of AA (0.01 mM – 1.0 mM). Means ± SEM of 3–10 experiments per group are shown. Cell volume is given in percent of the control value obtained during the last 15 minutes of the control period.

lar pH of C6 glioma cells was 7.3 during the control period (Table 1). AA led to a decrease of pH$_i$ in a dose-dependent manner. Upon administration of 0.05 mM, or 0.1 mM, pH$_i$ fell within 5 min to 7.12 ± 0.03 (p < 0.05), or to 7.06 ± 0.03 (p < 0.01), respectively. In experiments with 0.05 mM, pH$_i$ eventually recovered to the normal level, whereas pH$_i$ recovery remained incomplete when 0.1 mM were administered (Table 1). The AA-induced intracellular acidosis of glial cells could be significantly attenuated by the aminosteroid U-74389F (p < 0.01).

Lactacidosis

As in previous studies, the volume of astrocytes from primary culture increased immediately upon extracellular acidification, with the extent of cell swelling depending on the level of lactacidosis and the duration of exposure (Table 2) [19,35]. For instance, lactacidosis of pH 6.8 resulted in cell swelling of 105.0 ± 1.7 % of control after 10 minutes with a further increase during ongoing acidosis up to 113.1 ± 7.4 % after 55 minutes

(p < 0.01; Table 2). Lowering pH to 6.2 led to cell swelling of 110.0 ± 1.2 % after 10 minutes whereas of 121.4 ± 0.9 % within 55 minutes (p < 0.01). Following restoration of the medium pH to 7.4 after lactacidosis of pH 6.8, recovery of the glial cell volume was complete within 30 minutes, while after lactacidosis of pH 6.2 normalization of the cell volume was incomplete (Table 2). Under control conditions, the pH$_i$ of astrocytes was 7.06 ± 0.07, remaining constant for an observation period of two hours (Table 2). Upon extracellular acidification astrocytes were not capable to defend their normal pH$_i$. The intracellular pH fell to 6.70 ± 0.05, or 6.33 ± 0.05 within one minute during lactacidosis of pH 6.8 or 6.2, i.e. more or less down to the level of the medium pH. Reestablishment of the medium pH to 7.4 was followed by an immediate increase of pH$_i$ to approximately 7.15, a pH-level similar to that found before induction of lactacidosis in the suspension medium (Table 2).

Attenuation or almost complete inhibition of glial swelling from lactacidosis (at pH 6.2) was obtained by addition of amiloride or employment of a Na$^+$-free

Table 2. *Volume (Top) and pH$_i$ (Bottom) of Astrocytes from Primary Culture During Exposure to Lactacidosis*

Time [min]	Control, pH 7.4			Lactacidosis					pH 7.4			
	−15	−10	−5	1	5	10	30	55	61	65	70	90
pH 7.4	100.18 ±0.30	99.85 ±0.52	99.97 ±0.52	99.86 ±0.57	99.50 ±0.59	99.94 ±0.33	99.40 ±0.30	100.1 ±0.81	98.89 ±0.76	98.65 ±1.45	97.48 ±1.88	95.88 ±2.22
pH 6.8	100.13 ±0.13	100.03 ±0.18	99.85 ±0.24	103.58* ±0.54	105.84* ±1.42	105.03* ±1.74	109.62* ±2.64	113.11* ±7.44	101.87 ±5.83	102.98 ±5.99	100.24 ±7.85	96.25 ±8.81
pH 6.2	100.35 ±0.26	99.93 ±0.17	99.72 ±0.25	107.06* ±1.19	108.51* ±0.75	109.95* ±1.23	116.26* ±1.40	121.44* ±0.93	116.19* ±1.96	112.15* ±2.26	109.12* ±0.96	105.03* ±2.41
Time [min]	Control, pH 7.4			Lactacidosis					pH 7.4			
	−15	−10	−5	1	5	10	30	55	61	65	70	90
pH 7.4	7.05 ±0.06	7.06 ±0.07	7.08 ±0.07	7.08 ±0.07	7.09 ±0.07	7.10 ±0.08	7.08 ±0.07	7.10 ±0.07	7.09 ±0.07	7.08 ±0.07	7.09 ±0.07	7.10 ±0.07
pH 6.8	7.08 ±0.05	7.08 ±0.04	7.09 ±0.05	6.70* ±0.05	6.68* ±0.06	6.69* ±0.07	6.71* ±0.07	6.72* ±0.06	7.10 ±0.08	7.15 ±0.06	7.12 ±0.07	7.14 ±0.06
pH 6.2	7.05 ±0.03	7.07 ±0.05	7.07 ±0.04	6.33* ±0.05	6.25* ±0.03	6.20* ±0.10	6.25* ±0.04	6.27* ±0.03	7.05 ±0.04	7.13 ±0.02	7.14 ±0.03	7.15 ±0.02

* $p < 0.01$ vs. pH 7.4.
Cell volume (top) and intracellular pH (bottom) of astrocytes from primary culture after titration of the extracellular pH from 7.4 to 6.8, or 6.2 with isotonic lactic acid. 5–6 experiments were performed in each group. Cell volume and pH$_i$ following neutralization of the medium pH after 60 min exposure to lactacidosis is seen at the right.

suspension medium ($p < 0.01$). The intracellular acidosis of astrocytes, however, was more pronounced in these experiments, particularly in the studies using Na$^+$-free medium ($p < 0.01$). On the other hand, blocking of H$^+$-channels by ZnCl$_2$ not only was diminishing intracellular acidosis but also glial swelling from AA ($p < 0.01$).

As in previous studies, viability of astrocytes from primary culture was not affected by lactacidosis up to pH 6.2 within a time period of 2–3 hours, even not when inhibitors or modified media were employed [19,35].

Discussion

The present results demonstrate the powerful potential of AA or lactacidosis to induce swelling as well as intracellular acidosis of glial cells [19,35,36]. The effects were found at concentrations of AA or pH-levels, which occur in brain tissue in vivo under pathophysiological conditions, such as focal injury or ischemia [2,4,5,16,23,24].

Arachidonic Acid

Exposure of glial cells to abnormal levels of AA may initiate a variety of mechanisms which are likely to play a role in the cell swelling process. Metabolization of arachidonic acid along its various degradation

pathways with formation of prostaglandins, thromboxanes, hydroxy- or hydroperoxy fatty acids, leukotrienes and lipoperoxides or reactive oxygen radicals [41] altogether could be liable to contribute to cell swelling from arachidonic acid. Primary cultured astrocytes have been reported to generate superoxide radicals from AA [8]. The radicals in turn may lead to peroxidation of membrane lipids and accumulation of lipid peroxides by free radical attack and removal of H$^+$ from unsaturated fatty acids in a radical chain reaction [7]. Oxygen-derived free radicals and lipid peroxides not only damage plasma membranes but also membranes of subcellular organelles, such as mitochondria [8]. Damage of the cell membrane by these radical compounds is likely to increase its permeability to Na$^+$-ions as a mechanism underlying cell swelling. Moreover, the colloid osmotic pressure in the intracellular compartment caused by the negatively charged proteins might enhance the influx of Na$^+$-ions [39]. The resulting net accumulation of Na$^+$-ions in the cell cannot be compensated for by an increased activity of the Na$^+$-pump, because arachidonic acid is an inhibitor of the Na$^+$/K$^+$-ATPase [37]. Polyunsaturated fatty acids are known to have adverse effects on oxidative phosphorylation in mitochondria. Hillered and Chan found inhibition of the respiratory activity of isolated mitochondria from brain tissue by AA [12]. Inhibition of mitochondrial respiration is likely to activate glycolysis associated with an increased lactic acid production by the glial cells. As seen, administration of arachidonic acid to the glial cell suspension was lowering the intracellular pH in a dose-dependent fashion (Table 1). Whereas cell swelling was maintained for the remaining observation period, pH_i was found to recover within 50–80 min, though. Induction of intracellular acidosis by arachidonic acid might have resulted in a variety of responses, such as activation of H^+-exchange mechanisms and free radical reactions [11,33]. Enhancement of radical formation and of lipid peroxidation by acidosis has been observed in brain tissue homogenates. Generation of free radicals under these conditions specifically increases the protonated $\cdot O_2^-$ moiety, i.e. the level of hydroperoxyl radicals ($\cdot OOH$), which are stronger oxidants and more lipid soluble than superoxide radicals [33]. The maximum of intracellular acidification was attained within 5 min after addition of arachidonic acid to the glial cell suspension, however, followed by subsequent recovery of pH_i. Recovery of the intracellular pH was probably accomplished by activation of the Na^+/H^+-antiporter as a response to the decreasing pH_i [11]. The normalization of pH_i by the exchange of H^+- against Na^+-ions leading to intracellular accumulation of Na^+ may have contributed to the cell swelling from arachidonic acid [36].

Viability of the C6 glioma cells started to decline, when arachidonic acid concentrations in the suspension medium were raised to 0.5 mM, a level observed in brain tissue in vivo in cerebral ischemia[23]. The corresponding AA-concentration threshold associated with impairment of viability of primary cultured astrocytes was, however, definitely lower. This indicates that astrocytes were more vulnerable to arachidonic acid than glioma cells, irrespective of their lower viability (i.e. 80 %) already under control conditions [36]. The mechanisms underlying irreversible damage of astrocytes or of C6 glioma cells by arachidonic acid are not yet completely understood. Formation of oxygen derived free radicals and of lipid peroxides from the fatty acid must be considered among others. These reactive compounds are liable to damage almost all structural components of a cell, interfering with metabolism and function. Damage of the cell membrane may also result in an increased Ca^{2+}-permeability, permanently raising cytoplasmic Ca^{2+}-levels with its fatal consequences, such as mitochondrial Ca^{2+}-overload or activation of lipases and other catabolic enzymes [30]. The latter is considered to play a major role in neuronal and glial cell death in the brain, for example in cerebral ischemia [34]. Enhancement of Ca^{2+}-fluxes into synaptosomes of rat brain and spinal cord nerve cells by free radicals and lipid peroxidation products has been demonstrated in vivo [6]. The predominant role of increased cytoplasmic Ca^{2+}-levels notwithstanding, the pathochemical properties of arachidonic acid itself may suffice to inflict irreversible cell injury. Due to its amphophilic nature arachidonic acid can be incorporated into the plasma membrane [17]. At high concentrations, amphophilic fatty acids, such as arachidonic acid aggregate into micelles, acquiring the ability to incorporate membrane lipids [17]. Arachidonic acid containing micelles have detergent properties, rendering them capable of dissolving the plasma membranes.

Lactacidosis

The significant role of acidosis, especially lactacidosis, for swelling and damage of brain cells is known for quite a while [10,13,19,28,35]. The volume of C6 glioma cells as well as of astrocytes from primary culture was found to increase, when the extracellular pH was titrated to pH 6.8 or below (Table 2). Viability of the cells remained unchanged down to pH 6.2, however, at a pH-level of 5.6 the number of viable cells started to decline [35]. Concerning underlying mechanisms, glial swelling from acidosis has been assumed to be caused by membrane transport systems involved in pH_i-regulation, such as the Na^+/H^+-antiporter [11,26], Cl^-/HCO_3^--exchanger [22], and Na^+/HCO_3^--cotransporter [21,29]. Activation of these mechanisms as an attempt to defend pH_i during acidosis may finally lead to accumulation of Na^+- and Cl^--ions in the cell and, thereby, osmotically obliged water [13,19,35]. Accordingly, glial swelling from acidosis could be reduced or even completely blocked by amiloride, an inhibitor of the Na^+/H^+-antiporter or suspension of the cells in Na^+-free medium.

A most interesting observation in this study is probably that astrocytes from primary culture as well as C6 glioma cells were not able to maintain their normal pH_i during acidosis in the extracellular compartment, as also observed by other authors (Table 2) [26,28]. The results may indicate, that regulation of pH_i - although activated - simply failed, as it could not compensate the steep extra-/intracellular H^+-gradient induced by acidification of the extracellular compartment. This assumption is supported by augmentation of the intracellular acidosis during inhibition of the Na^+/H^+-antiporter by amiloride, or omission of Na^+-ions from the medium, respectively. Therefore, the Na^+-dependent transport systems seem to be activated for extrusion of H[+]-ions leading to accumulation of Na[+]-ions, hence, cell swelling, while they are not capable to control pH_i under these conditions [20,40]. The effect of $ZnCl_2$ is not completely explained so far, which was shown to reduce glial swelling and to attenuate intracellular acidosis at a medium pH of 6.2. A possible explanation is blocking of Zn^{2+}-sensitive H[+]-channels in glial cells, which have been demonstrated in other cell types, for instance snail neurons [25]. Flux of H[+]-ions trough these H[+]-channels among others may facilitate intracellular acidification upon induction of an extracellular acidosis.

The question remains, nevertheless, why the glial cells were not able to maintain their normal pH_i during extracellular acidosis, although pH_i-regulating mechanisms were likely to be activated. Obviously, a net intracellular acid load is dependent not only on extrusion of acid, but also on the velocity of its entry, as determined by H[+]-permeability of the cell membrane. In view of the fact, that glial cells have an important function for maintenance of homeostasis of the extracellular space in the brain, the lack of an effective pH_i-regulation might relate with their role in cerebral acid-base regulation. The polarity of glial cells, localized between neurons and the capillary bed could assign them to transport protons from the neuronal extracellular environment to the blood vessels. Such an unidirectional acid transport could be associated with a non-uniform distribution of transporters and ionchannels over the cell membrane in vivo [29]. In vitro, however, the polarity might disappear and transporters and ion-channels redistribute randomly over the cell membrane, making glial cells particularly vulnerable to acidification from environmental acidosis [20,40].

Taken together, the present data demonstrate that arachidonic acid as well as lactic acid induce swelling, intracellular acidosis, and damage of glial cells obtained from an established cell line or from primary culture. The findings support that release and accumulation of arachidonic acid and the development of tissue acidosis in the brain under pathological conditions, such as ischemia or trauma, play an important role in cytotoxic brain edema and cellular injury. The present results provide a basis for further studies on underlying mechanisms as well as for the development of specific methods of treatment.

Acknowledgements

The excellent technical and secretarial assistance of Ingrid Kölbl, Christa Grönlinger, and Helga Kleylein is gratefully acknowledged.

The study was supported by a grant of the Deutsche Forschungsgemeinschaft (Sta 406/2-1).

References

1. Aldrich EF, Eisenberg HM, Saydjari C, Luerssen TG, Foulkes MA, Jane JA, Marshall LF, Marmarou A, Young HF (1992) Diffuse brain swelling in severely head-injured children. A report from the NIH Traumatic Coma Data Bank. J Neurosurg 76: 450–454

2. Andersen BJ, Unterberg AW, Clarke GD, Marmarou A (1988) Effect of posttraumatic hypoventilation on cerebral energy metabolism. J Neurosurg 68: 601–607

3. Baethmann A (1978) Pathophysiological and pathochemical aspects of cerebral edema. Neurosurg Rev 1: 85–100

4. Baethmann A, Maier Hauff K, Schürer L, Lange M, Guggenbichler C, Vogt W, Jacob K, Kempski O (1989) Release of glutamate and of free fatty acids in vasogenic brain edema. J Neurosurg 70: 578–591

5. Bazan NGJ (1970) Effects of ischemia and electroconvulsive shock on free fatty acid pool in the brain. Biochim Biophys Acta 218: 1–10

6. Braughler JM, Duncan LA, Goodman T (1985) Calcium enhances in vitro free radical-induced damage to brain synaptosomes, mitochondria, and cultured spinal cord neurons. J Neurochem 45: 1288–1293

7. Braughler JM, Hall ED (1989) Central nervous system trauma and stroke. I. Biochemical considerations for oxygen radical formation and lipid peroxidation. Free Radic Biol Med 6: 289–301

8. Chan PH, Chen SF, Yu AC (1988) Induction of intracellular superoxide radical formation by arachidonic acid and by polyunsaturated fatty acids in primary astrocytic cultures. J Neurochem 50: 1185–1193

9. Frangakis MV, Kimelberg HK (1984) Dissociation of neonatal rat brain by dispase for preparation of primary astrocyte cultures. Neurochem Res 9: 1689–1698

10. Goldman SA, Pulsinelli WA, Clarke WY, Kraig RP, Plum F (1989) The effect of extracellular acidosis on neurons and glia in vitro. J Cereb Blood Flow Metab 9: 471–477

11. Grinstein S, Rothstein A (1986) Mechanisms of regulation of the Na[+]/H[+] exchanger. J Membr Biol 90: 1–12

12. Hillered L, Chan PH (1988) Effects of arachidonic acid on respiratory activities in isolated brain mitochondria. J Neurosci Res 19: 94–100

13. Jakubovicz DE, Klip A (1989) Lactic acid-induced swelling in C6 glial cells via Na[+]/H[+] exchange. Brain Res 485: 215–224

14. Kachel V, Glossner E, Kordwig E, Ruhenstroth Bauer G (1977) Fluvo-metricell, a combined cell volume and cell fluorescence analyzer. J Histochem Cytochem 25: 804–812

15. Kachel V (1986) Interactive multi-window integration of two-parameter flow cytometric data fields. Cytometry 7: 89–92

16. Katsura K, Ekholm A, Asplund B, Siesjö BK (1991) Extracellular pH in the brain during ischemia: relationship to the severity of lactic acidosis. J Cereb Blood Flow Metab 11: 597–599

17. Katz AM, Messineo FC (1981) Lipid-membrane interactions and the pathogenesis of ischemic damage in the myocardium. Circ Res 48: 1–16

18. Kempski O, Chaussy L, Groß U, Zimmer M, Baethmann A (1983) Volume regulation and metabolism of suspended C6 glioma cells: an in vitro model to study cytotoxic brain edema. Brain Res 279: 217–228

19. Kempski O, Staub F, Jansen M, Schödel F, Baethmann A (1988) Glial swelling during extracellular acidosis in vitro. Stroke 19: 385–392

20. Kempski O, Volk C (1994) Neuron-glia interaction during injury and edema of the CNS. In: Ito U, Baethmann A, Hossmann KA, Kuroiwa T, Marmarou A, Reulen HJ, Takakura K (eds) Brain edema IX. Acta Neurochir (Wien) [Suppl] 60: 7–11

21. Kettenmann H, Schlue WR (1988) Intracellular pH regulation in cultured mouse oligodendrocytes. J Physiol 406: 147–162

22. Kimelberg HK, Biddlecome S, Bourke RS (1979) SITS-inhibitable Cl⁻ transport and Na⁺-dependent H⁺ production in primary astroglial cultures. Brain Res 173: 111–124

23. Kinouchi H, Imaizumi S, Yoshimoto T, Motomiya M (1990) Phenytoin affects metabolism of free fatty acids and nucleotides in rat cerebral ischemia. Stroke 21: 1326–1332

24. Kraig RP, Pulsinelli WA, Plum F (1985) Heterogeneous distribution of hydrogen and bicarbonate ions during complete brain ischemia. In Kogure K, Hossmann KA, Siesjö BK, Welsh FA (eds) Progress in brain research, Vol 63. Elsevier, Amsterdam, pp 155–166

25. Mahaut-Smith M (1989) The effect of zink on calcium and hydrogen ion currents in intact snail neurons. J Exp Biol 145: 455–464

26. Mellergard PE, Siesjö BK (1991) Astrocytes fail to regulate intracellular pH at moderately reduced extracellular pH. Neuroreport 2: 695–698

27. Musgrove E, Rugg C, Hedley D (1986) Flow cytometric measurement of cytoplasmic pH: a critical evaluation of available fluorochromes. Cytometry 7: 347–355

28. Nedergaard M, Goldman SA, Desai S, Pulsinelli WA (1991) Acid-induced death in neurons and glia. J Neurosci 11: 2489–2497

29. Newman EA (1991) Sodium-bicarbonate cotransport in retinal Müller (glial) cells of the salamander. J Neurosci 11: 3972–3983

30. Orrenius S, McConkey DJ, Bellomo G, Nicotera P (1989) Role of Ca²⁺ in toxic cell killing. Trends Pharmacol Sci 10: 281–285

31. Rothe G, Valet G (1988) Phagocytosis, intracellular pH, and cell volume in the multifunctional analysis of granulocytes by flow cytometry 9: 316–324

32. Siesjö BK, Ingvar M, Westerberg E (1982) The influence of bicuculline-induced seizures on free fatty acid concentrations in cerebral cortex, hippocampus, and cerebellum. J Neurochem 39: 796–802

33. Siesjö BK, Bendek G, Koide T, Westerberg E, Wieloch T (1985) Influence of acidosis on lipid peroxidation in brain tissues in vitro. J Cereb Blood Flow Metab 5: 253–258

34. Siesjö BK, Bengtsson F (1989) Calcium fluxes, calcium antagonists, and calcium-related pathology in brain ischemia, hypoglycemia, and spreading depression: a unifying hypothesis. J Cereb Blood Flow Metab 9: 127–140

35. Staub F, Baethmann A, Peters J, Weigt H, Kempski O (1990) Effects of lactacidosis on glial cell volume and viability. J Cereb Blood Flow Metab 10: 866–876

36. Staub F, Winkler A, Peters J, Kempski O, Kachel V, Baethmann A (1994) Swelling, acidosis, and irreversible damage of glial cells from exposure to arachidonic acid in vitro. J Cereb Blood Flow Metab 14: 1030–1039

37. Swarts HG, Schuurmans Stekhoven FM, De Pont JJ (1990) Binding of unsaturated fatty acids to Na⁺, K⁺-ATPase leading to inhibition and inactivation. Biochim Biophys Acta 1024: 32–40

38. Thomas JA, Buchsbaum RN, Zimniak A, Racker E (1979) Intracellular pH measurements in Ehrlich ascites tumor cells utilizing spectroscopic probes generated in situ. Biochemistry 18: 2210–2218

39. Tomita M, Gotoh F (1992) Cascade of cell swelling: thermodynamic potential discharge of brain cells after membrane injury. Am J Physiol 262: H603–H610

40. Volk C, Haberstok J, Staub F, Klawe C, Peters J, Kempski O (1996) Acidosis and glial intracellular pH regulation. Submitted

41. Wolfe LS (1982) Eicosanoids: prostaglandins, thromboxanes, leukotrienes, and other derivatives of carbon-20 unsaturated fatty acids. J Neurochem 38: 1–14

Correspondence: Dr. Frank Staub, Klinik für Neurochirurgie, Universität zu Köln, Joseph-Stelzmann-Str. 9, D-50924 Köln, Federal Republic of Germany.

Acta Neurochir (1996) [Suppl] 66: 63–67
© Springer-Verlag 1996

Environmental Influence on Outcome After Experimental Brain Infarction

B. B. Johansson

Department of Neurology, Lund University Hospital, Lund, Sweden

Summary

After permanent ligation of the middle cerebral artery the motor function of rats housed in an enriched environment, i.e. cages with opportunities for various activities but not forcing the rats to do any particular task, is significantly better than in rats housed in individual cages. Rats kept in an enriched environment before and after MCA ligation improved sooner and slightly more than those placed in the enriched environment after ischemia but with no lasting significant difference except for climbing. Preliminary studies suggest that social stimulation is more important than physical activity. Rats with fetal neocortical grafts implanted into the infarct cavity performed better if exposed to enriched environment than grafted control rats housed in standard laboratory cages with 5 rats in each cage. However, they did not perform better than non-grafted rats housed in the same enriched environment. The infarct size did not differ between rats housed in an enriched environment and control rats. There was no correlation between infarct size and performance in rats exposed to an enriched environment. The improved motor function suggests that a rich environment may stimulate mechanisms that enhance brain plasticity.

Keywords: Experimental brain infarcts; environment; motor function; brain plasticity.

Introduction

Significant functional improvement occurs in most stroke survivors during the months following stroke onset [13,31,44]. To what extent specific rehabilitation can enhance the functional outcome is debated and no rehabilitation method has been shown to be superior to another [10,23,36,51]. Little attention has so far been given to the influence of environmental factors on functional outcome after focal brain ischemia. In the present report I will discuss some recent data on the behavioral effect of pre- and postoperative enrichment of the environment.

Material and Methods

In the first experiments the middle cerebral artery was ligated proximal to the striatal branches. The rats were either kept in individual cages before and after ligation of the right middle cerebral artery (MCA), transferred to an enriched environment after the ligation or kept in an enriched environment before as well as after the operation. The experimental protocol is described in detail elsewhere [35].

In a later experiment we studied if the enriched environment could enhance functional outcome in rats receiving neocortical transplants to the infarct cavity [19]. We then ligated the MCA distal to the striatal branches as described by Coyle [8] to induce an infarct limited to neocortex. Three weeks after the MCA occlusion fetal neocortical blocks of tissue were implanted into the infarct cavity in rats housed either in an enriched environment or in standard cages with five rats in each cage. A vehicle containing glucose 0.6 % and NaCl 0.9 % was deposited into the infarct cavity in control sham-transplanted rats housed in an enriched environment [19].

The enriched environment. The size of the cage was 815 mm × 610 mm × 450 mm. 150 mm above the floor, two horizontal boards, 70 mm wide, were placed along one of the sides. One board connected the floor with the elevated boards and, at a higher level still, one board was put across a corner. A chain, a swing, a swingboard and wooden blocks were placed in the box. Minor changes were made once a week adding new objects and withdrawing others.

Behavioral tests used include a postural reflex test, a limb placing test, beam-walking, traversing a rotating pole, climbing, inclined plane and paw-reaching. The tests have been described in detail elsewhere [17,19,35]. Since we present data from the limb placement test in the tables and figure 3, a summary of our modification [35] of the test described by De Ryck [9] is presented below.

In the 6 subtests, the placement of the forelimbs are studied and the hindlimbs are included in tests 4 and 6. During tests 1 through 4, the rat was held in a soft grip by the examiner. In test 1, limb placing was tested by slowly lowering the rat toward a table. At about 10 cm above the table, normal rats stretch and place both forepaws on the table. For test 2, with the rat's forelimbs touching the table edge, the head of the rats was moved 45° upward while the chin was supported to prevent the nose and the vibrissae from touching the table. (A rat with focal brain lesion may lose contact with the table with the paw contralateral to the injured hemisphere.) In test 3, forelimb placement of the rat when facing a table edge was observed. (A normal rat

64

B. B. Johansson

places both forepaws on the table top.) Test 4 recorded forelimb and hindlimb placement when the lateral side of the rat's body was moved toward the table edge. For test 5, the rat was placed on the table and gently pushed from behind toward the table edge. (A normal rat will grip on the edge, but an injured rat may drop the forelimb contralateral to the injured hemisphere.) Test 6 was the same as test 5 but the rat was pushed laterally toward the table edge. Each test was scored as follows: 0, no placing; 1, incomplete and/or delayed (>2 seconds) placing; and 2, immediate and correct placing. For each body side, the maximum score from the tests used was 16 = normal behavior.

Determination of infarct volume. At the termination of the experiments, i.e. 13–14 weeks after operation, when a cystic cavity is formed, the remaining area of the right hemisphere was determined in pixel size and was subtracted from the contralateral hemispheric area. The volume was calculated from the section thickness and section frequency. At this late stage after an MCA occlusion, the infarct volume may be overestimated by the inclusion of secondary tissue losses such as thalamic atrophy. To eliminate any error due to hemispheric volume losses, the total tissue volume loss is expressed in per cent of contralateral intact hemisphere, the cortical tissue loss in per cent of contralateral cortex and the thalamus in per cent of contralateral thalamus [19,35].

Statistics. For tests based on scoring systems (ordinal measures) the Kruskal Wallis non-parametric analysis of variance test (ANOVA) was used with a multiple comparison post hoc test to determine the number and relation of the group differences at 95 % significance level. For difference in infarct volume and the paw-reaching test, one-way parametric ANOVA with Scheffé's post hoc procedure was used at 95 % significance levels.

Results

In the leg placement test, rats housed in single cages had significantly lower scores than the other groups at all times (Table 1). Rats housed in an enriched environment both before and after the operation had higher scores, i.e. performed significantly better than rats placed in the enriched environment after the operation only at five weeks. The discriminative values of the limb placement subtests varied with time. In subtest B, when visual stimuli and whisker contact with the sur-

Table 1. *Mean Neurological Scores for the Left Paw in a Leg Placement Test after Ligation of the Right Middle Cerebral Artery in Spontaneously Hypertensive Rats*

| Time after operation | Groups | | |
	A n=9	B n=10	C n=12
2 weeks	3.9 ± 1.8[a]	7.3 ± 2.0	8.8 ± 2.4
5 weeks	6.9 ± 2.3[b]	8.1 ± 2.6[b]	11.2 ± 2.4
7 weeks	8.1 ± 1.8[a]	11.6 ± 1.2	12.9 ± 2.4

Mean ± SD. Score 16 = normal behavior.
A rats housed in individual cages; *B* rats placed in an enriched environment after the operation; *C* rats kept in an enriched environment before and after the operation.
The significant group differences for the left paws were p = 0.002, 0.007 and 0.003 at 2,5 and 7 weeks;[a] significantly different from B and C; [b] significantly different from C (Kruskal-Wallis non-parametric ANOVA and a multiple comparison post hoc test both at 95 % significance level).

Table 2. *p-Values for Group Differences in the Subtests A-F in Leg Placement at Various Times after Ligation of the Middle Cerebral Artery*

		2 weeks	5 weeks	7 weeks
A	Forelimb	0.0025	0.0049	–
B	Forelimb	0.0036	0.0028	0.0001
C	Forelimb	–	0.04	0.013
D	Forelimb	0.0023	–	–
	Hindlimb	–	–	0.0245
E	Forelimb	–	–	–
F	Forelimb	–	–	–
	Hindlimb	0.0102	–	–

Kruskal-Wallis non-parametric ANOVA at 95 % significance level.

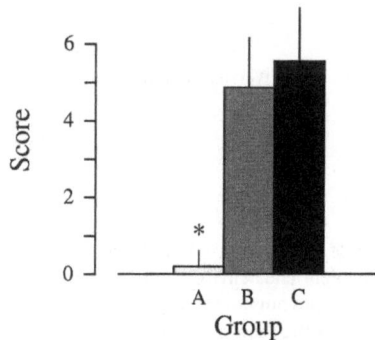

Fig. 1. Scores for rats traversing a non-rotating pole 10 weeks after a proximal MCA occlusion (cortical and striatal infarct). Mean values ± SD. *A* rats housed in individual cages; *B* rats placed in an enriched environment after the operation; *C* rats kept in an enriched environment before and after the operation. Based on data from Ohlsson and Johansson [35]. Score 6 = the rat traverses the pole without any problem

face are prevented, the difference was significant at all times tested and increased with time (Table 2). Figure 1 shows the results from traversing a non-rotating pole 10 weeks after the MCA occlusion. Neither the infarct volume, nor the thalamic atrophy differed between the groups (Fig. 2).

Grafted rats housed in an enriched environment performed significantly better than grafted rats in standard laboratory cages but not better than non-grafted rats in an enriched environment (Figs. 3 and 4). In none of the studies did the enriched environment enhance the performance in the paw-reaching test, a test for skilled forelimb use.

Discussion

Our data show that an enriched postoperative environment can significantly improve function after focal brain ischemia. To what extent improved functional outcome is due to recovery of lost functions or to compensation for lost functions is difficult to ascertain

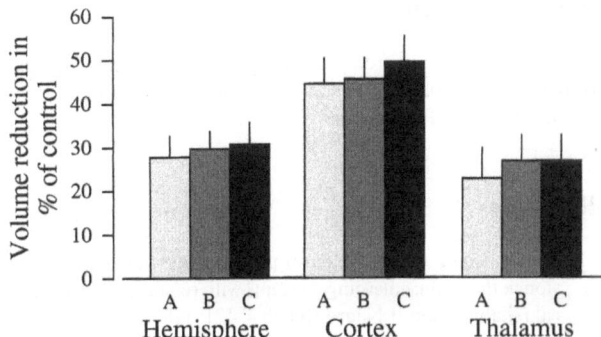

Fig. 2. Brain tissue volume reduction in per cent of contralateral hemisphere, cortex and thalamus 13 weeks after proximal MCA occlusion. Based on data from Ohlsson and Johansson [35]

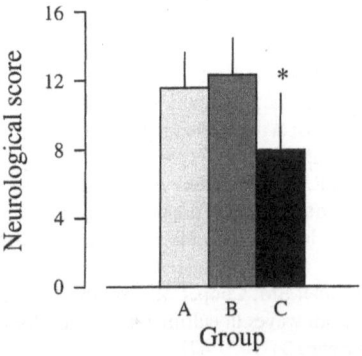

Fig. 3. Leg placement test 9 weeks after distal MCA occlusion resulting in a pure cortical infarct. *Group A* housed in enriched environment; *Group B* housed in an enriched environment and fetal neocortical tissue implantation 3 weeks after MCA occlusion; *Group C* fetal, cortical tissue implanted 3 weeks after MCA occlusion. Rats were housed in standard 5-rat cages. Group C is significantly different from group A and B. Based on data published in Grabowski *et al.* [19]. Score 16 = no deficits

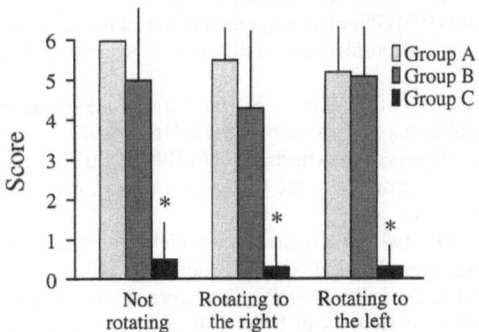

Fig. 4. Scores on traversing a rotating or a non-rotating pole 9 weeks after distal MCA occlusion. Groups as in Fig. 3. Based on data from Grabowski *et al.* [19] Score 6 = the rat traverses the pole without any problem

[29,30,42,55]. In agreement with earlier studies, the performance in the paw-reaching test changed little with time [17,35]. The fact that we, in spite of afferent and efferent connections with the host brain [18,19], in

contrast to Kelche *et al.* [28] could see no further improvement in grafted rats might be due to the complexity of the focal brain infarct model or to the late time of transplantation [19].

Mechanisms that have been proposed to account for functional improvement after stroke include resolution of brain edema, absorption of necrotic tissue and disappearance of remote functional depression or "diaschisis" [12]. Several factors might be involved but current data suggest that a substantial part of functional recovery after stroke might be attributed to brain plasticity [1,5,20,24,26,27,53,54].

The plasticity of the adult brain may be larger than previously expected. A number of elegant studies have shown that the afferent input to the motor, somatosensory, auditory, or visual cortex can alter the cortical topography [3,4,6,20–22,33,37–40,47]. That our cortical maps are changed by experience throughout life has been shown to occur, e.g. in skill acquisition, after repeated electrical stimulation to the cortex and after amputation. The question is to what extent it occurs after lesions like brain infarcts, and if so, at what level of the central nervous system and what is the time course.

It seems likely that some reorganization of brain functions can take place after focal brain infarction. Based on activation studies in patients who have recovered function after a subcortical stroke [5,53,54] it has been proposed that recruitment of cortical areas in the non-damaged hemisphere as well as adjacent to the lesion may be important for functional recovery. Whether this is due to "unmasking" of existing networks or establishing of new networks is not known, nor is it known if the increased blood flow necessarily signifies improved function. Furthermore, there is evidence for substantial difference in individual activation pattern after stroke recovery [54]. To firmly establish a correlation between the metabolic activation and functional recovery, studies comparing the degree in pattern of activation in patients with good or less good recovery are essential. Whether or not subcortical structures and cerebellum can compensate for lost cortical functions remains to be established.

There is experimental evidence for a specific period after the brain damage during which the behavioral experience can maximally modify the neural events. Schallert and Jones have shown that there is a substantial transient expansion of the identical cortical area of the opposite hemisphere after a unilateral injury to the forelimb representation area. Expansion was greatest in regions containing large pyramidal neurons in layer

5, the major output neurons of the cerebral cortex. The time-dependent dramatic increase in the complexity and extent of dendritic arborization was greatest during the second and third postoperative weeks and had not returned to control levels at 120 days [26]. Activity was necessary for the changes to take place. Thus, if the movement of the intact forelimb was restricted, there was no increased arborization on the side opposite the lesion. [27].

It has been hypothesized that mechanisms involved in restoration of function after brain lesions may parallel mechanisms in ontogenesis and in learning [6,21,22,24,27,33]. In addition to neuronal mechanisms, the possible role of astrocytes [7,14,32,50] and of nonsynaptic transmission[1] is discussed. Networks of astrocytes may provide an extraneuronal system for rapid long-distance signaling [7,32,50]. Evidence for nonsynaptic transmission has been extensively reviewed by Bach-y-Rita [1].

An enriched environment can increase protein content, dendritic branching and number of dendritic spines per unit length of dendrite and may stimulate transmitters and nerve growth factors (for ref. see [24,28–30,34,41,43,55]). Environmental stimuli capable of inducing synaptogenesis result in hypertrophy and increased number of astrocytes [48].

If functional improvement after focal brain ischemia is related to brain plasticity, drugs taken during the post ischemic event may be crucial for the outcome [11,12,16,25,45,46,49,52]. Transmitters including noradrenaline, glutamate, acetylcholine, serotonin and dopamine as well as certain peptides, hormones and trophic factors may be involved for events promoting brain plasticity. Of particular interest is that drugs that are currently tested for possible neuroprotection in the acute phase such as glutamate antagonists could have a negative effect if given after the acute stage. Other drugs that on the basis of experimental data have been proposed to inhibit brain plasticity include α-adrenergic antagonists, phenytoin and GABA agonists [2,11,12,16,25,45]. In a retrospective analysis of the functional outcome in stroke patients to durgs taken at stroke onset, some evidence was obtained that this could be the case also in man [15]. However, more data are needed in this clinically highly relevant area. To what extent enriched environment and drugs interact is another area that remains to be explored [52].

Acknowledgements

Supported by grants from the Swedish Medical Research Council (project 14X-4968), King Gustaf V and Queen Victoria Foundation and the Segerfalk Foundation.

References

1. Bach-y-Rita P (1990) Receptor plasticity and volume transmission in the brain: emerging concepts with reference to neurological rehabilitation. J Neuro Rehab 4: 121–128
2. Brailowsky S, Knight RT, Blood K, Scabini D (1986) γ-aminobutyric acid-induced potentiation of cortical hemiplegia. Brain Res 362: 322–330
3. Calford MB, Tweedale R (1990) Interhemispheric transfer of plasticity in the cerebral cortex. Science 249: 805–807
4. Chino YM, Kaas JH, Smith III EL, Langston AL, Cheng H (1992) Rapid reorganization of cortical maps in adult cats following restricted deafferentation in retina. Vision Res 32: 789–796
5. Chollet F, DiPiero V, Wise RJS, Brook DJ, Dolan RJ, Frackowiak RSJ (1991) The functional anatomy of motor recovery after stroke in humans: a study with positron emission tomography. Ann Neurol 29: 63–71
6. Cohen LG, Brasil-Neto JP, Pascual-Leone A, Hallett M (1993) Plasticity of cortical motor output organization following deafferentation, cerebral lesions, and skill acquisition. Adv Neurol 63: 187–200
7. Cornell-Bell AH, Finkbeiner SM, Cooper MS, Smith SJ (1990) Glutamate induces calcium waves in cultured astrocytes: long-range glial signaling. Science 247: 470–473
8. Coyle P (1982) Middle cerebral artery occlusion in the young rat. Stroke 13: 855–859
9. De Ryck M, Van Reempts J, Borgers M, Wauquier A, Janssen AJ (1989) Photochemical stroke model: flunarizine prevents sensorimotor deficits after neocortical infarcts in rats. Stroke 20: 1383–1389
10. Ernst E (1990) A review of stroke rehabilitation and physiotherapy. Stroke 21: 1081–1085
11. Feeney DM, Gonzalez A, Law WA (1982) Amphetamine, haloperidole and experience interact to affect the rate of recovery after motor cortex injuries. Science 217: 855–857
12. Feeney DM (1991) Pharmacologic modulation of recovery after brain injury: a reconsideration of diaschisis. J Neurol Rehab 5: 113–128
13. Fries W, Danek A, Scheidtmann K, Hamburger C (1993) Motor recovery following capsular stroke. Brain 116: 369–382
14. Gage FH, Olejniczak P, Armstrong DM (1988) Astrocytes are important for sprouting in the septohippocampal circuit. Exp Neurol 102: 2–13
15. Goldstein LB (1991) Pharmacologic modulation of recovery after stroke: clinical data. J Neurol Rehab 5:129–140
16. Goldstein LB, Davis JN (1990) Post-lesion practice and amphetamine-facilitated recovery of beam-walking in the rat. Restor Neurol Neurosci 1: 311–314
17. Grabowski M, Brundin P, Johansson BB (1993) Paw-reaching, sensorimotor, and rotational behavior after brain infarction in rats. Stroke 24: 889–895
18. Grabowski M, Brundin P, Johansson BB (1993) Functional integration of cortical grafts placed in brain infarcts of rats. Ann Neurol 34: 362–368

19. Grabowski M, Sørensen J C, Mattson B, Zimmer J, Johansson BB (1995) Influence of an enriched environment and cortical grafting on functional outcome in brain infarcts of adult rats. Exp Neurol 133: 1–7

20. Jenkins WM, Merzenich MM (1987) Reorganization of neocortical representations after brain injury: a neurophysiological model of the bases of recovery from stroke. Prog Brain Res 71: 249–266

21. Jenkins WM, Merzenich MM, Ochs MT, Allard T, Guíc-Robles E (1990) Functional reorganization of primary somatosensory cortex in adult owl monkeys after behaviorally controlled tactile stimulation. J Neurophysiol 63: 82–104

22. Jenkins WM, Merzenich MM, Recanzone G (1990) Neocortical representational dynamics in adult primates: implications for neuropsychology. Neuropsychologia 28: 573–584

23. Johansson BB (1993) Has sensory stimulation a role in stroke rehabilitation ? Scand J Rehab Med [Suppl] 29: 87– 96

24. Johansson BB, Grabowski M (1994) Functional recovery after brain infarction: Plasticity and neural transplantation. Brain Pathol 4: 85–95

25. Johansson BB (1995) Functional recovery after brain infarction. A review of animal data. Cerebrovasc Dis 5: 278–281

26. Jones TA, Schallert T (1992) Overgrowth and pruning of dendrites in adult rats recovering from neocortical damage. Brain Res 581: 156–160

27. Jones TA, Schallert T (1994) Use-dependent growth of pyramidal neurons after neocortical damage. J Neurosci 14: 2140–2152

28. Kelche C, Dalrymple-Alford JC, Will B (1988) Housing conditions modulate the effects of intracerebral grafts in rats with brain lesions. Behav Brain Res 28: 287–295

29. Kolb B (1992) Mechanisms underlying recovery from cortical injury: reflections on progress and directions for the future. In: Rose FD, Johnson DA (eds) Recovery from brain damage. Plenum, New York, pp 169–186

30. Kolb B, Gibb R (1991) Environmental enrichment and cortical injury: behavioral and anatomical consequences of frontal cortex lesions. Cereb Cortex 1: 189–198

31. Kotila M, Waltimo O, Niemi M-J, Laaksonen R, Lempinen M (1984) The profile of recovery from stroke and factors influencing outcome. Stroke 15: 1039–1044

32. Marrero H, Astion ML, Coles JA, Orkand RK (1989) Facilitation of voltage-gated ion channels in frog neuroglia by nerve impulses. Nature 339: 378–380

33. Merzenich MM, Recanzone G, Jenkins WM, Allard TT, Nudo RJ (1988) Cortical representational plasticity. In: Rakic P, Singer W (eds) Neurobiology of neocortex. Wiley, Berlin, pp 42–67

34. Mohammed AK, Winblad B, Ebendal T, Lärkfors L (1990) Environmental influence on behaviour and nerve growth factor in the brain. Brain Res 528: 62–72

35. Ohlsson A-L, Johansson BB (1995) The environment influences functional outcome of cerebral infarction in rats. Stroke 26: 644–649

36. Ottenbacher KJ, Jannell S (1993) The results of clinical trials in stroke rehabilitation research. Arch Neurol 50: 37–44

37. Pascual-Leone A, Grafman J, Hallett M (1994) Modulation of cortical motor output maps during development of implicit and explicit knowledge. Science 263: 1287–1289

38. Pascual-Leone A, Torres F (1993) Plasticity of the sensorimotor cortex representation of the readingfiger in Braille readers. Brain 116: 39–52

39. Pons TP, Garraghty PE, Mishkin M (1988) Lesion-induced plasticity in the second somatosensory cortex of adult macaques. Proc Natl Acad Sci USA 85: 5279–5281

40. Recanzone GH, Schreiner CE, Merzenich MM (1993) Plasticity in the frequency representation of primary auditory cortex following discrimination training in adult owl monkeys. J Neurosci 13: 87–103

41. Rose FD (1988) Environmental enrichment and recovery of function following brain damage in the rat. Med Sci Res 16: 257–263

42. Rose FD, Johnson DA (1992) Recovery from brain damage. Adv Exp Med Biol 325: 187–198

43. Rosenzweig MR (1984) Experience, memory, and the brain. Am Psychol 39: 365–376

44. Skilbeck CE, Wade DT, Langton Hewer R, Wood VA (1983) Recovery after stroke. J Neurol Neurosurg Psychiatry 14: 5–8

45. Schallert T, Hernandez TD, Barth TM (1986) Recovery of function after brain damage: severe and chronic disruption by diazepam. Brain Res 379: 104–111

46. Schallert T, Jones T, Weaver M, Shapior L, Crippens D, Fulton R (1992) Pharmacologic and anatomic considerations in recovery of function. In: Hanson S, Tucker DM (eds) Neuropsychological assessment - physical medicine and rehabilitation: state of the art reviews 6: 375–393

47. Sanes JN, Wang J, Donoghue JP (1992) Immediate and delayed changes of rat motor cortical output representation with new forelimb configurations. Cereb Cortex 2: 141–152

48. Sirevaag AM, Greenough WT (1991) Plasticity of GFAP-immunoreactive astrocyte size and number in visual cortex of rats reared in complex environments. Brain Res 540: 273–278

49. Sutton RL, Feeney DM (1992) α-Noradrenergic agonists and antagonists affect recovery and maintenance of beam-walking ability after senorimotor cortex ablation in the rat. Rest Neurol Neurosci 4: 1–11

50. Usowicz MM, Gallo V, Cull–Candy SG (1989) Multiple conductance channels in type-2 cerebellar astrocytes activated by excitatory amino acids. Nature 339: 380–383

51. Wagennar RC, Meijer OG (1991) Effects of stroke rehabilitation (1). A critical review of the literature. J Rehab Sciences 4: 61–73

52. Walsh R (1981) Sensory environments, brain damage, and drugs: a review of interactions and mediating mechanisms. Int J Neurosci 14: 129–137

53. Weiller C, Chollet F, Friston KJ, Wise RJS, Frackowiak RSJ (1992) Functional reorganization of the brain in recovery from striatocapsular infarction in man. Ann Neurol 31: 463–472

54. Weiller C, Ramsay SC, Wise RJS, Friston KJ, Frackowiak RSJ (1993) Individual patterns of functional reorganization in the human cerebral cortex after capsular infarction. Ann Neurol 33: 181–189

55. Will B, Kelche C (1992) Environmental approaches to recovery of function from brain damage: a review of animal studies (1981–1991). Adv Exp Med Biol 325: 79–103

Correspondence: Barbro Johansson, M.D., Department of Neurology, Lund University Hospital, S-221 85 Lund, Sweden.

Acta Neurochir (1996) [Suppl] 66: 68–72

Fetal Neocortical Grafts Placed in Brain Infarcts Do Not Improve Paw-Reaching Deficits in Adult Spontaneously Hypertensive Rats

M. Grabowski[1], **B. B. Johansson**[1], and **P. Brundin**[2]

[1]Department of Neurology, University Hospital and [2]Department of Medical Cell Research, University of Lund, Sweden

Summary

The aim was to study if neural grafts placed in brain infarcts could improve functional recovery.

The middle cerebral artery was occluded (MCAO) in 19 spontaneously hypertensive rats. Nine rats were sham operated. Twelve to 16 days after the ischemic insult, 9 of the MCAO rats received transplants of dissociated fetal neocortical tissue (MCAO-T) and 1, 3 and 6 months after transplantation surgery, the rats were behaviorally evaluated by a test for forelimb function. Infarct and transplant sizes were measured morphometrically.

The remaining volume of the infarcted hemisphere was 66 ± 7 % (mean ± SD) in the MCAO group and 71 ± 9 % in the MCAO-T group of the non-operated hemisphere. All grafted rats had surviving transplants. Contralateral to the lesion, paw-reaching was highly impaired in both infarcted groups compared with sham-operated controls with no significant difference between MCAO and MCAO-T. The lesion size correlated significantly with contralateral paw-reach performance at all test periods.

We conclude that neocortical grafts did not alleviate the impaired forepaw function.

Keywords: Cerebral ischemia; transplantation; neocortex; behavior.

Introduction

The high incidence of stroke and resulting neurological deficits in humans initiated our interest in developing a transplantation model in focal brain ischemia, induced by a permanent occlusion of the middle cerebral artery (MCAO) [19]. Preparation of fetal neocortex for grafting follows the original description by Björklund *et al.* [1]. Since hypertension is the most important risk factor for cerebrovascular disease in the adult population, we use adult hypertensive rats as recipients of the fetal tissue. Using this model we have shown that fetal neocortical grafts survive when placed in the infarcted area [9]. Good survival is obtained with tissue within a wide range of fetal development. A time-delay between the insult the grafting surgery is beneficial for graft survival and the host brain environment seems to be most hospitable around 3 weeks after arterial occlusion. The grafts consist of tissue of a neuronal appearance with pyramidal-like cells but lack the lamination of normal neocortex. The grafts are vascularized by leptomeningeal vessels and a capillary network which forms a plexus on the brain surface. Grafts have a lower density of capillaries than typically found in neocortex, but the morphology of capillaries is normal [6]. The host innervates the grafts by cholinergic, noradrenergic and serotonergic fiber systems. Ingrowth of afferent fibers from the host thalamus and neocortex also occurs [5]. Transplanted neurons develop an extensive axonal network within the grafts but efferent connections with the host are rare [10]. Graft glucose metabolism is increased following stimulation of the host somatosensory pathway which demonstrates that transplanted neurons can be functionally integrated with neural circuitries of the host [8].

Behavioral effects of cortical grafts after brain infarction have not yet been reported. Therefore, the present study was designed to evaluate whether neocortical grafts ameliorate the deficits of forelimb function that result from a middle cerebral artery occlusion in the rat [7].

Material and Methods

Twenty-eight adult male inbred spontaneously hypertensive rats (SHR-Mol) (Møllegaard Breeding & Research Centre, Denmark), weighing 200–220 g at the start of the experiment, were used in the study. Fetal donor tissue was obtained by caesarian section from pregnant females of the same strain from the same supplier.

Ketamine 50–100 mg/kg (Ketalar, Parke-Davis) and xylazine 5–10 mg/kg (Rompun, Bayer) was used for MCAO and graft surgery. Methohexital 100–120 mg/kg (Brietal, Lilly) preceded intracardiac perfusion. All anesthetics were given i.p.

Nineteen rats were subjected to a right MCAO by ligating the vessel with a 10–0 monofilament nylon suture (Deknatel, Germany) at the point where it crosses the olfactory tract [19]. Nine control rats were sham operated by means of an incision of the skin and the underlying fascia of the temporal muscle, followed by suturing.

Nine rats were grafted with fetal neocortex 12–16 days after MCAO. Six fetuses of 17–18 days of gestation (crown-rump length 24–25 mm) were obtained from pregnant rats. According to previous descriptions [2,9], a strip of neocortical anlagen was dissected bilaterally from each fetus and dissociated to a final volume of 200 μl. Using a 10 μl Hamilton microsyringe, 1.5 μl of the dissociated tissue was injected stereotactically in the infarcted right hemisphere at six different sites in each animal, according to the following coordinates (given in mm, with the toothbar set at –3.3): (1) A: 2.7, L: 3.0, V: 2.5; (2) A: 1.7, L: 2.5, V: 2.5; (3) A: 1.7, L: 5.0, V: 6.0; (4) A: 1.7, L: 5.0, V: 3.0; (5) A: –3.3, L: 3.0, V: 2.5; (6) A: –3.3, L: 6.3, V: 6.0. The anterior and lateral coordinates are with reference to bregma, whereas the ventral coordinates are with reference to the surface of the skull at bregma.

One, 3 and 6 months after transplantation surgery, the rats were behaviorally evaluated by a test for forelimb function. This test measures separately the reaching and grasping capacity of the right and left forelimb as previously described [7,14]. Briefly, the rat was placed in a plexiglas box which contains a central evaluated platform with a staircase on each side. The staircases have six steps, each baited with eight 45-mg chow pellets (Campden Instruments, England), making a total of 48 pellets on each side. The rat is placed on the platform and may from this position collect pellets by mouth or tongue from the top two steps. The ipsilateral forepaw must be used to reach pellets from the lower steps. Limb function was estimated from the number of pellets eaten on each side. The rats were tested daily for 9 days a month after surgery, repeated for 3 days after 3 and 6 months. Each test period was preceded by a 48-hour period of food deprivation, resulting in about a 10 % reduction of body weight. One test lasting 20 minutes was performed daily, after which the rats were transferred to their home cages and given approximately 20 g of standard food pellets to maintain body weight.

After completion of behavioral testing, all lesioned rats were perfused via the ascending aorta with 0.9 % saline followed by a fixative containing 4 % formaldehyde in a 0.1 mol/L phosphate buffer. The brains were postfixed overnight and then stored in 20 % sucrose in 0.1 mol/L phosphate buffer until sectioning. The brains were cut in 40-μm-thick coronal sections on a freezing microtome. At 200-μm intervals, two series were retrieved for cresyl violet staining and acetylcholine-esterase (AChE) histochemistry [13].

The size of the each hemisphere was calculated by measuring the cross-sectional area at 12 coronal levels (approximately 1 mm between each measurement) starting 3.7 mm anterior to bregma [15]. The volume was achieved from the cross-sectional areas and the distance between the sections. The graft volume was determined from area measurements of every section stained for cresyl violet. To obtain cross-sectional areas the software Image Grabber 2.03 (Neotech, USA) and Image/MG 1.44ß (National Institutes of Health, USA) were used with an image analyzing system which consisted of a video camera (Dage MTI, USA), a light box (Imaging Research, Canada) and a Macintosh IIsi computer.

Results

Infarct and Graft Morphology

One rat in the MCAO group and 3 rats in the MCAO-T group died during the experimental period. The other rats showed cystic brain infarcts affecting the neocortex and lateral part of caudate putamen. The ipsilateral thalamus was atrophic. The ischemic damage resulted in a volume reduction of the operated hemisphere which did not differ significantly between the groups (P = 0.27, unpaired T-test). The remaining volume was 66 ± 7 % (mean ± 1 SD) in the MCAO group and 71 ± 9 % in the MCAO-T group of the non-operated hemisphere.

All rats in the MCAO-T group had surviving grafts with volumes varying between 2.7–23.3 mm^3. The grafts formed lobular masses in the infarcted area (Fig. 1) but single injections of neural tissue were misplaced in two rats so that a small portion of the transplant was found in normal brain or on the brain surface. Graft morphology was similar to what has been described in previous studies [5,9]. Microscopically, the tissue was composed of neurons and glial cells, often arranged in lobulated structures separated by thin septa. The laminar organization of normal neocortex was lacking. A similar density of AChE-positive nerve fibers was seen in all grafts. The transplants were attached to the host neocortex or striatum and the host-graft border was often demarcated by a thin gliotic zone.

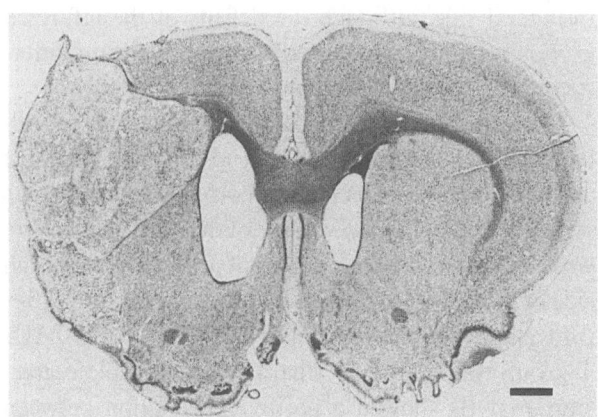

Fig. 1. Photomicrograph of a coronal section showing a brain six months after receiving implantation of dissociated fetal neocortical tissue into the infarcted area after middle cerebral artery occlusion. The infarct cavity is completely occupied by the graft which is arranged in lobules. The graft lacks the lamination of normal neocortex. Scale bar is 1 mm. Cresyl violet stain

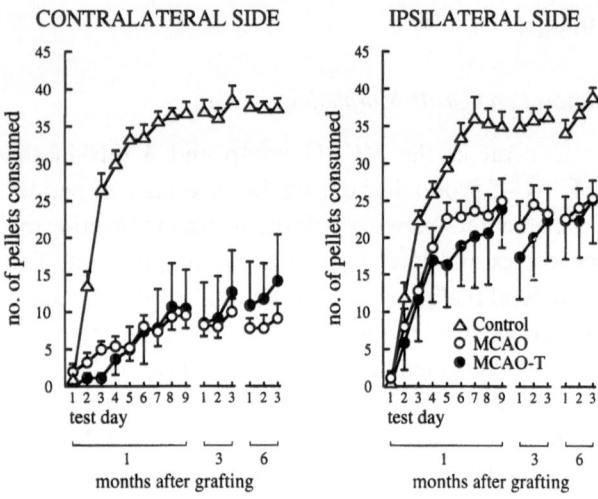

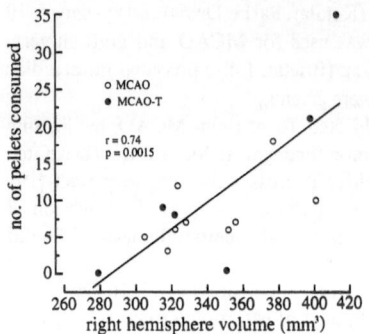

CONTRALATERAL SIDE IPSILATERAL SIDE

Fig. 3. Correlation analysis between lesion size and contralateral paw-reach performance (means of test days 1 to 3) 6 months after middle cerebral artery occlusion (*MCAO*) or MCAO followed by transplantation (*MCAO-T*)

Fig. 2. Line graphs of results of paw-reach test, showing number of consumed pellets ipsilaterally and contralaterally to sham operation, middle cerebral artery occlusion *(MCAO)* or MCAO followed by transplantation (*MCAO-T*). Performance did not differ significantly between MCAO and MCAO-T. Values are means and error bar is 1 SEM

Table 1. *Correlation Coefficients Between Lesion Size and Outcome of the Paw-Reach Test on the Ipsilateral (Operated) and Contralateral Side for the Combined MCAO and MCAO-T Group*

	1 Month	P	3 Months	P	6 Months	P
Ipsilateral	0.78	0.0007	0.73	0.002	0.72	0.0027
Contralateral	0.83	0.0001	0.75	0.0012	0.74	0.0015

The following measures were used for the analysis: lesion size (ipsilateral hemisphere volume), paw-reach test (mean performance during days 7–9 and days 1–3).

Discussion

In agreement with a previous study [7], we have confirmed that brain infarction after MCAO produces impairments of both contralateral and ipsilateral forepaw function. The deficits are stable over time and show a correlation to lesion size.

The dissociated fetal neocortical tissue which was transplanted into brain infarcts did not influence behavioral outcome as measured by the paw-reaching test. The failure of functional recovery could have several explanations. Although the grafts are richly innervated by the host [5], axonal growth from transplant to host is sparse [10]. Functional recovery by reconstruction of normal pathways is therefore limited which may explain the present results.

Cortical grafts [12,16,18] or infusion of neurotrophic factors [4,17,21] may provide support for intrinsic host neurons which have lost their normal targets provided that intervention is performed immediately after the cortical lesion. Although it remains to be shown, it is possible that one of these procedures could ameliorate functional deficits by rescuing neurons projecting to the neocortex. Such a trophic effect is unlikely in the present experiment because grafting was performed about 2 weeks after MCAO. One reason for choosing

Paw-Reaching

Figure 2 shows the number of pellets consumed on the contralateral and ipsilateral side for all groups in the paw-reach test. On the contralateral side, an analysis of variance (ANOVA) from month 1 to 6 yielded a significant group difference (P=0.0001) which reflects the precise grasp ability of the control group. The MCAO and MCAO-T groups displayed large paw-reaching deficits and post-hoc comparisons (Scheffé's test) did not disclose significant differences between the groups on any test day. Similar results were seen on the ipsilateral side, although the deficits of the infarcted groups were less pronounced with no significant differences between the MCAO and MCAO-T group.

The extent of ischemic brain damage was significantly correlated to the outcome of the paw-reach test at all test periods. The highest correlation coefficients were achieved when the volume of the right hemisphere was compared with the number of pellets eaten on each side of the staircase (Table 1). The result 6 months after transplantation is plotted in Fig. 3. When the MCAO-T group was examined separately, a multiple regression analysis showed a positive correlation between contralateral forepaw function and right hemisphere volume (standard coefficient 0.72, P=0.014), whereas a negative correlation was found to graft volume (standard coefficient –0.5, P=0.037). No significant relationship was found between right hemisphere volume and graft volume (r=0.24, P=0.65).

this late time is that graft survival is poor in our model when transplanting immediately after the insult [9].

As already mentioned, the grafts are innervated by several fiber systems which normally project to the neocortex [5]. This is a requirement for an integration of neural circuitries but is not necessarily beneficial for functional recovery after an ischemic insult. In this study increasing graft volume significantly correlated to poorer paw-reach performance. It may be argued that the host innervation of the graft could interfere with the neural reorganization occurring after the insult and possibly compromise the spontaneous functional improvement which is often seen in clinical and experimental stroke. However, the main result of the present study does not indicate harmful effects of the grafts because behavioral outcome did not differ between the grafted and non-grafted group.

Graft-induced functional recovery has previously been demonstrated after precise lesions of a neural pathway or anatomical area associated to a specific behavior, e.g. the nigro-striatal dopamine system and the hippocampus. Our transplantation paradigm and application of the paw-reach test differs in several respects from models. Although previous results [7] suggested that the paw-reach test was suitable for functional evaluation after MCAO over a long period of time, problems are associated with its application to the MCAO transplantation model. The first problem is related to the lesion, which is large and by no means restricted to the cortical region where the forepaw is somatotopically organized. Although the stereotactic procedure for implantation is quite precise, the ultimate placement of the graft is unpredictable, because the injections are made into a large necrotic mass or into a liquid-filled infarct cyst. Possibly the paw-reach test would get a "fairer trial" in a different grafting model (e.g. placement of grafts into circumscribed cortical lesions of the forepaw somatotopical area). The second problem associated with the paw-reach test and the MCAO transplantation model is that paw-reaching is dependent on cortical *and* striatal function [20]. In view of the striatum's general importance for motor function in the rat, the appropriate experiment in the present model might therefore be grafting of striatal, rather than cortical tissue to the infarcted area.

In summary, MCAO is attractive because it resembles clinical ischemic stroke but may not be optimal in a transplantation paradigm as used in the present study. To that end, we recently modified our model by performing MCAO distal to the origin of the lenticulostriate branches [3] which results in a neocortical infarction without affecting the caudate putamen. This enables us to study neocortical deficits and behavioral effects of neocortical grafts without the influence of striatal damage. Further, the use of block transplants instead of dissociated tissue seems to be favorable in reconstruction of host-graft neural circuitries [11].

References

1. Björklund A, Stenevi U, Schmidt RH, Dunnett SB, Gage FH (1983) Intracerebral grafting of neuronal cell suspensions I. Introduction and general methods of preparation. Acta Physiol Scand [Suppl] 522: 1–7
2. Brundin P, Strecker RE (1991) Preparation and intracerebral grafting of dissociated fetal brain tissue in rats. In: Conn PM (ed) Methods in neurosciences, Vol 7. Lesions and transplantation. Academic Press, San Diego, pp 305–326
3. Coyle P (1982) Middle cerebral artery occlusion in the young rat. Stroke 13: 855–859
4. Eagleson KL, Cunningham TJ, Haun F (1992) Rescue of both rapidly and slowly degenerating neurons in the dorsal lateral geniculate nucleus of adult rats by a cortically derived neuron survival factor. Exp Neurol 116: 156–162
5. Grabowski M, Brundin P, Johansson BB (1992) Fetal neocortical grafts implanted in adult hypertensive rats with cortical infarcts following a middle cerebral artery occlusion: ingrowth of afferent fibers from the host brain. Exp Neurol 116: 105–121
6. Grabowski M, Christofferson RH, Brundin P, Johansson BB (1992) Vascularization of fetal neocortical grafts implanted in brain infarcts in spontaneously hypertensive rats. Neuroscience 51: 673–682
7. Grabowski M, Brundin P, Johansson BB (1993) Paw-reaching, sensorimotor, and rotational behavior after brain infarction in rats. Stroke 24: 889–895
8. Grabowski M, Brundin P, Johansson BB (1993) Functional integration of cortical grafts placed in brain infarcts of rats. Ann Neurol 34: 362–368
9. Grabowski M, Johansson BB, Brundin P (1994) Survival of fetal neocortical grafts implanted in brain infarcts of adult rats: the influence of post-lesion time and the age of donor tissue. Exp Neurol 127: 126–136
10. Grabowski M, Johansson BB, Brundin P (1995) Neocortical grafts placed in the infarcted brain of adult rats: few or no efferent fibers grow from transplant to host. Exp Neurol 134: 273–276
11. Grabowski M, Sørensen JC, Mattsson B, Zimmer J, Johansson BB (1995) Influence of an enriched environment and cortical grafting in brain infarcts of adult rats. Exp Neurol 133: 96–102
12. Haun F, Cunningham TJ (1984) Cortical transplants reveal CNS trophic interactions in situ. Dev Brain Res 15: 290–294
13. Hedreen JC, Bacon SJ, Price DL (1985) A modified histochemical technique to visualize acetylcholinesterase-containing axons. J Histochem Cytochem 30: 134–140
14. Montoya CP, Campbell-Hope LJ, Pemberton KD, Dunnett SB (1991) The "staircase test": a measure of independent forelimb reaching and grasping abilities in rats. J Neurosci Meth 36: 219–228
15. Paxinos G, Watson C (1982) The rat brain in stereotaxic coordinates. Academic Press, Sydney

16. Sharp FR, Gonzalez MF (1986) Fetal cortical transplants ame- liorate thalamic atrophy ipsilateral to neonatal frontal cortex lesions. Neurosci Lett 71: 247–251
17. Sofroniew MV, Isacson O, Björklund A (1986) Cortical grafts prevent atrophy of cholinergic basal nucleus neurons induced by excitotoxic cortical damage. Brain Res 378: 409–415
18. Sørensen JC, Zimmer J, Castro AJ (1989) Fetal cortical trans- plants reduce the thalamic atrophy induced by frontal cortical lesions in newborn rats. Neurosci Lett 98: 33–38
19. Tamura A, Graham DI, McCulloch J, Teasdale GM (1981) Focal cerebral ischaemia in the rat: 1. Description of tech- nique and early neuropathological consequences following middle cerebral artery occlusion. J Cereb Blood Flow Metab 1: 53–60
20. Whishaw IQ, O'Connor WT, Dunnett SB (1986) The contribu- tions of motor cortex, nigrostriatal dopamine and caudate- putamen to skilled forelimb use in the rat. Brain 109: 805–843
21. Yamada K, Kinoshita A, Kohmura E, Sakaguchi T, Taguchi J, Kataoka K, Hayakawa T (1991) Basic fibroblast growth factor prevents thalamic degeneration after cortical infarction. J Cereb Blood Flow Metab 11: 472–478

Correspondence: Martin Grabowski, M.D., Department of Neurology, University Hospital, S-221 85 Lund, Sweden.

Acta Neurochir (1996) [Suppl] 66: 73–75
© Springer-Verlag 1996

Extended Studies on the Effect of Glutamate Antagonists on Ischemic CA-1 Damage

N. H. Diemer, T. Balchen, T. Bruhn, T. Christensen, I. Vanicky, M. Nielsen, and F. F. Johansen

Institute of Neuropathology, Molecular Neuropathology Unit, University of Copenhagen, Denmark

Summary

Glutamate receptors are numerous on the ischemia vulnerable CA-1 pyramidal cells. Postischemic use of the AMPA antagonist NBQX has shown up to 80 % protection against cell death. Three aspects of this were studied: In the first study, male Wistar rats were given NBQX (30 mg/kg × 3) either 20 hours or immediately (0 h) before 12 min of 4-vessel occlusion with hypotension. After six days of reperfusion comparison with an untreated group showed almost full protection in the 0 h group (4 % cell loss, $p < 0.001$) but only slight protection in the 20 h group (62 % cell loss, $p < 0.05$).

After 12 min of ischemia in the present model, eosinophilic CA-1 cells are seen from day 2 on. Since there could be a late, deleterious calcium influx via NMDA receptors, one group of ischemic rats was given MK-801 (5 mg/kg i.p.) 24 hours after ischemia. However, quantitation 6 days later of remaining CA-1 cells showed no protection.

In the third study referred here, two groups of ischemic rats were given NBQX (30 mg/kg × 3) immediately after ischemia. The groups survived for six and 21 days, respectively. Countings of CA-1 pyramidal cells showed an equal, significant protection in both groups (approx 20 % cell loss).

Keywords: 4-vessel occlusion; NBQX; MK-801.

Introduction

Several classes of drugs have shown to offer protection against global cerebral ischemia induced loss of hippocampal pyramidal CA1 neurons. Until now, glutamate antagonists of the AMPA type have turned out to be most effectively protective. One common mechanism of protection could be that the various drugs are changing the balance between excitation and inhibition and thereby lowering the energy metabolism stress during the sensitized state of the neuron in the postischemic period. Since the prevalent excitatory receptors on the ischemia vulnerable CA1 neurons (as well as the number of other vulnerable neuron populations) are of the glutamate type, this can explain why AMPA antagonists are so effective in protecting these particular neurons (Sheardown *et al.* 1990, Diemer *et al.* 1990)

However, in most studies glutamate antagonist have been given immediately before or after ischemia and the survival period has seldomly exceeded 1 week.

Thus, in the present investigation we studied the effect of preischemic administration of an AMPA antagonist (2,3-Dihydroxy-6-nitro-7-sulfamoylbenzo-(F)quinoxaline, NBQX) as well as the long term effect of a protective dose of NBQX, given immediately post-ischemically. Finally the effect of late administration of an NMDA antagonist (dizocilpine, MK-801) was evaluated.

Material and Methods

Male Wistar rats (250–300 g) were subjected to transient forebrain ischemia using a modification of Pulsinelli and Brierleys four-vessel occlusion model.

Six groups of animals were established; (8 animals in each)

1) rats given NBQX (30 mg/kg × 3 i.p.) 20 hours before ischemia; 6 days survival;
2) rats given NBQX (30 mg/kg × 3 i.p.) immediately before ischemia; 6 days survival;
3) rats given NBQX (30 mg/kg × 3 i.p.) immediately after ischemia; 6 days survival;
4) rats given NBQX (30 mg/kg × 3 i.p.) immediately after ischemia; 21 days survival;
5) rats given MK-801 (5 mg/kg i.p.) 24 hours after ischemia; 6 days survival;
6) Vehicle treated rats, 6 days survival.

Both vertebral arteries were electrocoagulated in methohexital (50 mg/kg i.p.) anaesthesia. The rats were allowed to recover with free access to water but were fasted overnight. The following day the rats were anaesthetized with 1 % halothane in a 2:1 N_2O/O_2 mixture, intubated and mechanically normoventilated with a rodent respirator (New England Medical Instruments Inc) The rats had a femoral artery cannulated to record mean arterial blood pressure (MABP) and to obtain blood samples for measurements of PO₂, PCO₂, pH and plasma glucose concentration (ABL 2 acid-base laboratory, Radiometer Copenhagen and Beckman Glucose Analyzer 2). The common carotid arteries were gently exposed and 3–0 silk ligatures threaded through polyethylene tubing were placed around them. The halothane was turned off and before the rats recovered from anaesthesia the carotids were occluded by tightening the ligatures and blood was withdrawn from the femoral artery until the pupils dilated and became unresponsive to light, while keeping the mean arterial blood pressure at 60 mm Hg. The body and head (temporal muscle) temperature was kept at 37–37.5 °C and 36 °C, respectively. After 12 min the ligatures were released and the blood reinfused. The animals were allowed to survive according to the schedule shown above.

Quantitations

The perfusion fixed brains were cut subserially in 7 μm frontal sections. Determination of cell density in the intermediate zone of the dorsal hippocampus was performed using an ocular grid and × 40 magnification on 4 sections. Statistical analysis was performed using Kruskal-Wallis one-way analysis of variance by ranks, and a significance level of 0.05 was chosen.

Results

The median values of the quantitations of CA-1 cells in the dorsal hippocampus are summarized in Fig. 1. NBQX given immediately before or after ischemia offered the best protection as quantitated after 6 days reperfusion. Treatment with NBQX 20 hrs before ischemia resulted in a moderate protection (62 % CA-1 neruones lost).

Treatment with MK-801 24 hours after ischemia and 6 days reperfusion showed no protection.

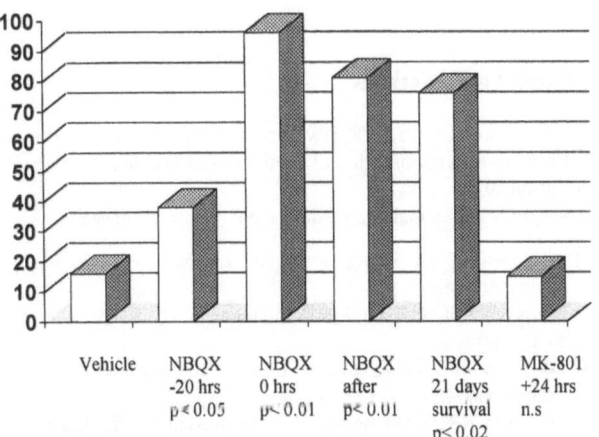

Fig. 1. Neurone density in hippocampal CA-1 zone. Percent remaining CA-1 neurons in dorsal hippocampus in the various groups outlined in the material and methods section

Discussion

NBQX Treatment

The intracellular mechanisms of AMPA antagonist protection against ischemia induced damage have not been disclosed, but an unspecific reduction of the excitation/inhibition ratio leading to a reduced energy metabolism stress is a likely possibility. Other classes of compounds have also shown (somewhat less) protective effects, which can be explained by such a common mechanism. Since some of the most sensitive neuron types also have a high density of glutamate receptors, this might explain the protective effect of e.g. AMPA antagonists after global ischemia.

NBQX, which is a competitive AMPA antagonist has been found to induce longterm effects on metabolic processes like protein synthesis (Frank *et al.* 1992). If this effect is essential, the possibility exists that even administration of NBQX several hours before ischemia could offer protection. The present study confirm such an effect, although it is weak. Whether a long term reduction of glucose metabolism is also involved is not known.

MK-801 Treatment

In contrast to the striking effect of NMDA antagonists in the reduction of infarct size after middle cerebral artery occlusion (Gill *et al.* 1991) the vast majority of studies of global ischemia have shown no protective effect of NMDA antagonists (i.e. MK-801, given shortly after ischemia, Diemer *et al.* 1990). One important finding in rats with MCAO is the uncoupling of glucose metabolism and blood flow in the infarct border zone (Nedergaard *et al.* 1988) as shown by double tracer autoradiography. As demonstrated by Mies *et al.* (1993) the spreading depressions (or anoxic depolarisations) which are responsible for the increased glucose metabolism in the infarct border zone are not – as normally – accompanied by flow increases. After global ischemia; except for a short, initial period with increased glucose (Diemer and Siemkowicz 1982) there is a long lasting reduction of both blood flow and metabolism in the hippocampus (Pulsinelli *et al.* 1982) without signs of repeated spreading depressions (Suzuki *et al.* 1983). However several hours after ischemia two important observations have been made, which could involve activation of NMDA receptors. Firstly, Andiné *et al.* (1991) found increases in extracellular glutamate and aspartate, 480 min after ischemia and secondly, Silver and Erecinska (1992) found at a similar reperfusion time, an increase in intracellular free calcium concentration. Since the NMDA receptor is gating a calcium permeable channel, the neuronal death in CA1 which mainly becomes visible at day 2 could involve a late pathological activation of this channel. The present study showed, however, that even after late administration of MK-801 there is no protective effect of this drug. Thus it is still unlikely that NMDA receptor mediated mechanisms are operative in the production of delayed neuronal death in CA1.

Long Term Effect of NBQX Treatment

The present study showed that in rats which survived for 21 days after 12 min of 4-VOH, the same degree of protection was found as in rats surviving for 6 days. In a study with 7 or 28 days survival after 10 min of 4-VO and NBQX treatment there were 45 % injured CA-1 neurons at day 7 and 89 % injured neurons at day 28 (Li and Buchan 1995). This significant difference indicates that with this paradigm, the NBQX-induced protection is not permanent. The results of the two studies cannot be directly compared since the models differ sligthly as well as the length of ischemia and survival periods. Clearly, it is important to study even longer survival periods, and in cases with only temporary portection, the effect of long-term treatment with AMPA antagonists (and other classes of compounds) must be investigated.

Acknowledgements

This study was supported by grants from the Danish Health Science Council and the Danish State Biotechnological Programme (PharmaBiotec). Eva Rahtkens is thanked for excellent technical support.

References

1. Andiné P, Orwar O, Jacobson I, Sandberg M, Hagberg H (1991) Changes in extracellular amino acids and spontaneous neuronal activity during ischemia and extended reflow in the CA1 of the rat hippocampus. J Neurochem 57: 222–229
2. Diemer NH, Siemkowicz E (1980) Increased 2-deoxyglucose uptake in hippocampus, substantia nigra and globus pallidus after cerebral ischemia. Acta Neurol Scand 61: 56–63
3. Diemer NH, Johansen FF, Jørgensen MB (1990) N-methyl-D-aspartate and non-N-methyl-D-aspartate antagonists in global cerebral ischemia. Stroke 21: [Suppl 11]: III 39–42
4. Frank L, Bruhn T, Diemer NH (1992) Regional cerebral protein synthesis after transient ischemia in the rat: Effect of the AMPA antagonist NBQX. Neurosci Lett 140: 108–112
5. Gill R, Brazell C, Woodruf GN (1988) The neuroprotective action of dizocilpine (MK-801) in the rat middle cerebral artery occlusion model of focal ischemia. Br J Pharmacol 103: 2030–2036
6. Li H, Buchan AM (1995) Progressive loss of hippocampal CA1 neurons following transient forebrain ischemia. J Cereb Blood Flow Metab 15 [Suppl 1]: S246
7. Nedergaard M, Jakobsen J, Diemer NH (1988) Autoradiographic determination of cerebral glucose content, blood flow, and glucose utilization in focal ischemia of the rat brain: influence of the plasma glucose concentration. J Cereb Blood Flow Metab 8: 100–108
8. Mies G, Kohno K, Hossmann KA (1993) MK-801, a glutamate antagonist, lowers flow threshold for inhibition of protein synthesis after middle cerebral artery occlusion of rat. Neurosci Lett 155: 65–68
9. Pulsinelli WA, Levy DE, Duffy TE (1982) Regional cerebral blood flow and glucose metabolism following transient forebrain ischemia. Ann Neurol 11: 499–509
10. Sheardown MJ, Nielsen EØ, Hansen AJ, Jacobsen P, Honoré T (1990) 2,3-Dihydroxy-6-nitro-7-sulfamoylbenzo(F)quinoxaline: a neuroprotectant for cerebral ischemia. Science 247: 571–574
11. Silver IA, Erecinska M (1992) Ion homeostasis in rat brain in vivo: intra- and extracellular [Ca^{2+}] and [H$^+$] in hippocampus during recovery from short-term, transient ischemia. J Cereb Blood Flow Metab 12: 759–772
12. Suzuki R, Yamaguchi T, Li C-L, Klatzo I (1983) The effects of 5-minute ischemia in Mongolian gerbils. II. Changes in spontaneous activity in cerebral cortex and CA1 sector of hippocampus Acta Neuropathol 60: 217–222

Correspondence: N. H. Diemer, M.D., Institute of Neuropathology, Molecular Neuropathology Unit, University of Copenhagen, 11 Frederik V's Vej, DK-2100 Copenhagen, Denmark.

Acta Neurochir (1996) [Suppl] 66: 76–80
© Springer-Verlag 1996

Thrombolysis in Acute Ischemic Stroke

C. Fieschi, C. Cavalletti, D. Toni, M. Fiorelli, M. L. Sacchetti, M. De Michele, M. C. Gori, E. Montinaro, and **C. Argentino**

Department of Neurological Sciences, University of Rome "La Sapienza", Rome, Italy

Summary

Thrombolysis is an attractive but potentially dangerous therapy for cerebral ischemia: it is capable of dissolving an arterial thrombus, but can also transform a pale infarct into a hematoma and/or may cause severe oedema and herniation. The safety and efficacy of the treatment critically depend on the timing of intervention and on patient selection.

In recent studies on ischemic stroke, spontaneous hemorrhagic transformation of an infarct seems to be related to the size of the lesion, and can be reliably predicted as early as five hours from stroke onset by the presence of focal hypodensity in the CT scan. That is why in the European Co-operative Acute Stroke Study (ECASS), a randomised, double blind trial on intravenous rt-PA in hemispheric stroke, patients showing, on the admission CT scan, extended early hypodensity, involving more than one third of the territory of the middle cerebral artery, were excluded from the study.

Other ongoing trials on thrombolytic agents are expected to provide further indications on how to identify those patients most likely to benefit and least likely to experience adverse effects from this treatment.

Keywords: Thrombolysis; stroke; reperfusion; hemorrhagic infarction.

Introduction

Experimental models of focal cerebral ischemia and studies conducted with positron emission tomography on ischemic stroke patients show that neuronal death can be prevented by restoring the cerebral blood flow to physiological values within a certain time interval, called the "therapeutic window". Restoring blood flow to the ischemic area is therefore the most important therapeutic goal in the acute phase of stroke (the other possible approach being to protect neurons by manipulating the adverse biochemical environment created by the ischemic insult). At present the therapeutic window in man is thought to be very similar to that of primates (about 6–8 hours) [21,22,34].

Thrombolytic agents are drugs with proven ability to dissolve an arterial thrombus that have occasionally been used for acute stroke since the late 50s [28,35,37]. After initial enthusiasm, this therapeutic strategy was abandoned because of the excessive risk of causing cerebral hemorrhage [29]. It must be remembered that early trials were conducted before the introduction of CT scanning, and consequently that some patients with intracerebral hemorrhage may have been treated. Moreover, patients were included up to many hours after clinical onset of stroke. These early experiences supported the idea that thrombolysis was contraindicated in acute cerebral ischemia.

The success of coronary thrombolysis [11,39] and the diffusion of modern diagnostic tools (CT scan and MRI), together with greater knowledge of the pathophysiology of brain ischemia, have renewed the interest in thrombolytic therapy for acute ischemic stroke. Thrombolytic drugs have thus recently been tested in some controlled studies. It is not yet possible to definitively say what the risk/benefit balance of thrombolytic therapy in acute ischemic stroke is. This work, based on medical literature and on our own experience, is a contribution to the discussion on critical topics, such as hemorrhagic transformation and brain oedema.

Spontaneous Hemorrhagic Transformation of Brain Infarct-Incidence, Causes, Consequences

Hemorrhagic transformation (HT) is a well-known occurrence in the evolution of a brain infarct [27,37], which led some early investigators to consider pale and hemorrhagic infarcts to be different moments of the same event [9,10,12,27]. The real incidence of HT is difficult to asses, and percentages may vary greatly in different reported series [18,19,27,38] owing to the fact that there are few prospective studies which use serial CT scanning at predetermined time intervals after acute ischemic stroke [4,13,18,31], and that the criteria of case selection in published reports are not homogeneous (autoptic studies, retrospective series, prospective CT studies in which only patients showing clinical deterioration were scanned). Moreover, some of the patients in some series were given antiplatelet agents or anticoagulant therapies, which might have influenced the incidence of bleeding.

In discussing these data it is important to distinguish between hemorrhagic infarction (HI), in which the brain elements in the infarcted area are interspersed with blood, and parenchymal hematoma (PH), which is a lake of blood destroying brain parenchyma. While PH is a feared event associated with clinical worsening, the same cannot always be said for HI, which is provocatively considered by some authors to be only a marker of a large lesion, not exerting *per se* a negative influence on the clinical outcome [33].

The global incidence of HT ranges from 15 to 45 % in different papers, PH accounting for about 5 % [18,27,37]. We recently analysed the data of a series of 150 continuous patients (unpublished data) with a first-ever ischemic supratentorial stroke, diagnosed on the basis of the clinical picture and of a CT scan performed within five hours of onset. A repeat CT scan was performed between day 5 and day 9 after clinical onset in all surviving patients (the 7 patients who died before the second CT scan underwent autopsy). HT was observed in 65 (43 %) patients. Of these, 58 had HI, while 5 (8 %) had a small and 2 (3 %) a massive PH.

The factors promoting HT are still a matter of discussion [15]. Bleeding within the infarcted area has a peak during the first week after stroke [30], and its occurrence or worsening has been attributed to hypertension, vasodilation after ischemia, reperfusion, concomitant medications (antithrombotic and thrombolytic agents) and aetiology of ischemia, since HT is more frequently detected in patients with cardioembolic stroke [26,32].

There is now a consensus that HT is more frequent in patients more severely affected at hospital admission [18,31] and with larger infarcts detected on CT scan [7,25,31,32]. Some recent work has focused attention on the relationship between early focal hypodensity on the CT scan performed in the first hours after stroke and subsequent HT [3,5,14,16,36] (Figs. 1,2).

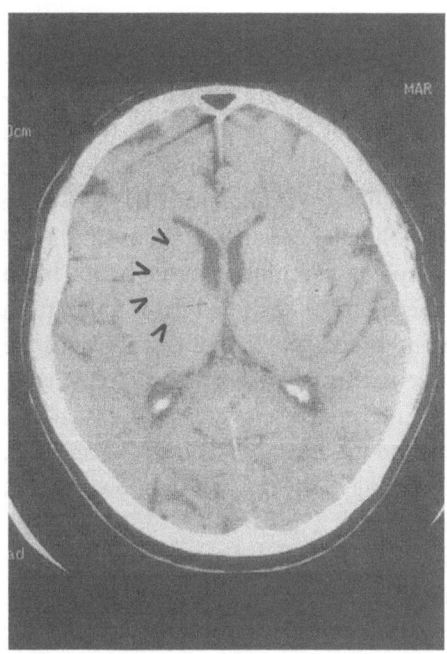

Fig. 1. First CT scan performed at 4 hours of stroke onset. Subcortical early focal hypodensity: obscuration of left lenticular nucleus

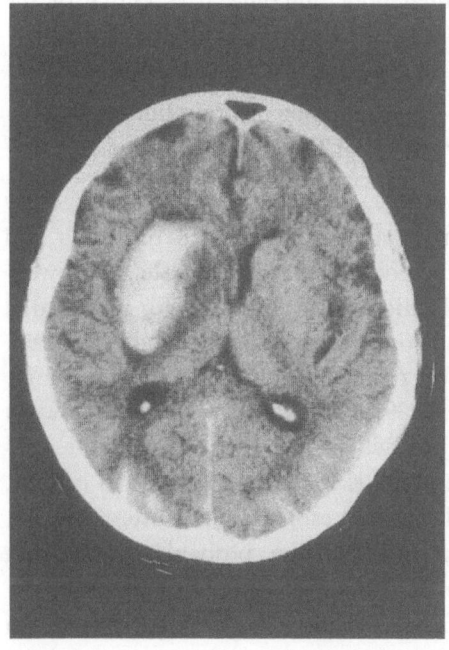

Fig. 2. Same patient, control CT scan at 5 days: parenchymal hematoma in the same territory of the early focal hypodensity in Fig. 1

In a classical paper of 1951 [10], Fisher and Adams first postulated that the restoration of blood flow (following spontaneous thrombolysis and/or distal clot migration) into a vascular bed weakened by ischemia could be responsible for HT. Recent studies [2,19,27] which have addressed this problem concluded that only some HTs are caused by the re-opening of large arterial vessels. According to our data (above mentioned, unpublished series) this mechanism could account for about 10 % of all bleedings, i.e. those with characteristics of PH. In the remaining patients the most probable mechanism leading to HT seems to be the reopening of the pial collaterals, following a decrease in the compression initially exerted by the brain oedema [18]. The spotty or scattered petechial pattern of bleeding, with cortico-subcortical distribution, the presence in these patients of early focal hypodensity, which has been interpreted as a sign of intracellular oedema [19,36], and the delay between stroke and HT, usually corresponding to the resolution of oedema, all support this hypothesis.

Apart from the infrequent cases of massive hematoma, HT does not seem to influence prognosis, whereas large oedematous infarcts do appear to be correlated with a poor outcome in HT patients [33]. According to our data (previously cited, unpublished series), the larger the lesion or the more severe the mass effect, the higher the frequency both of early deterioration and of final poor outcome, irrespective of the presence of HT.

Hemorrhagic Transformation of Brain Infarct in Patients Treated with Thrombolytic Drugs

Thrombolytic agents given for any reason obviously increase the risk of bleeding both in the brain and elsewhere in the body. Cerebral hemorrhage is a well-known complication of thrombolysis for myocardial infarction, and concomitant brain ischemia is, in fact, a criterion for patient exclusion from cardiologic protocols [11,20,39]. However, in a recent review on thrombolysis for acute ishemic stroke [37], the incidence of symptomatic PH was estimated to be 5 %, which is similar to that estimated for natural history. This apparent contradiction probably arises from differences in the timing of drug administration: in trials on myocardial infarction, as in early trials on cerebral ischemia, thrombolytic treatment was started many hours or even days after stroke onset [29,39]. It therefore appears to be the timing as opposed to the recanalization, that plays a key role in HT [27,37]. Lyden and Zivin in a recent paper hypothesised that very early recanalization may protect ischemic brain from reperfusion-mediated hemorrhage, whereas late treatment may promote this event [27].

In view of such considerations, several double blind, controlled thrombolysis trials were recently been started [6,8,17,23].

MAST-E is a multicenter, double-blind, controlled trial on streptokinase in stroke, which was recently interrupted, after the enrollment of 270 patients, because the incidence of cerebral hemorrhage was 2.8 times higher in the treated group. In-hospital mortality of treated patients was more than twice that of the control group and the mortality rate at 6 months was 33 % higher. According to the authors, their unfavorable results might be attributable to the severity of the stroke cases included. In any case, these data clearly indicate that, for the time being, the use of thrombolytic agents should be reserved for patients taking part in clinical trials in which safety is carefully monitored.

Our group participated in the ECASS (European Co-operative Acute Stroke Study), a randomised, double blind trial on intravenous rt-PA (Alteplase, 1.1 mg/Kg bw) given within 6 hours of the onset of a hemispheric stroke [23]. Patients with severe neurological deficit and/or presenting on the admission CT scan early sings of a large infarct (more than one third of the territory of middle cerebral artery) were excluded from the study. Recruitment was completed in spring 1994. Six hundred and twenty patients were randomised and included in the intention-to-treat analysis. Statistical analysis of all the data has not been completed yet, but it will be interesting to verify the safety of t-PA in this cohort of less severe strokes.

The Problem of Severe Oedema and Brain Herniation in Thrombolysed Patients

Severe infarct oedema is reported to occur spontaneously in approximately 5 % of large infarcts [26,37]. Although the cause is not clear, it may, it is hypothesised, be due to "reperfusion injury", and if so, thrombolysis could worsen the condition [24]. Wardlaw et al. in 1992 prospectively examined reperfusion in patients with MCA territory infarcts and found that infarct swelling was greatest in the largest infarcts and in patients showing permanent MCA occlusion at angiography or transcranial doppler ultrasound, whereas patients who reperfused within the first 3 days had smaller infarcts, less oedema and a better clinical outcome [38].

The frequency of severe swelling and brain herniation varies in different thrombolysis studies: some of them do not take account of this event as a possible complication [37], others report an incidence comparable to that of natural evolution. Alberts et al. recently reported on the incidence of oedema in 87 acute ischemic stroke patients treated with IV rt-PA, examined by means of angiography [1]. They found moderate or severe oedema in 28 % of the cases. Oedema was significantly associated with large vessel occlusions and large infarct volumes, but not with immediate recanalization status, blood pressure, blood glucose levels, or used diuretics. In the previously-mentioned MAST-E study [17] the proportion of deaths definitely known to be due to massive cerebral oedema was the same in the treated and placebo groups (26.5 %). As for HT, the problem of severe oedema seems to be strictly connected with the size of the lesion, it being more frequent in large infarcts.

Final Considerations

There is evidence that thrombolytic agents can be beneficial in acute ischemic stroke. Their impact in this disease will probably not, however, be as great as in acute myocardial infarction, owing to the higher frequency of life-threatening complications and the consequent need for more restrictive patient selection criteria. Rather than giving a definitive answer on the role of thrombolysis in stroke therapy, the results of the ongoing trials will probably provide further information on how to identify those patients who are most likely to benefit and least likely to experience adverse effects from thrombolytic therapy.

Note added in proof. Since the preparation of this manuscript, both of the above mentioned ongoing studies have been published [2,3]. We support the editorial comment by del Zoppo [1]: "The NINDS study and the ECASS experience are important steps toward effective treatment for stroke-related injury, but in the clinical application of thrombolysis in acute stroke we should move forward cautiously and with clear awareness of potential risks".

1. del Zoppo GJ (1995) Acute stroke-on the threshold of a therapy? N Engl J Med 333: 1632–1633
2. Hacke W, Kaste M, Fieschi C *et al.* (1995) Intravenous thrombolysis with recombinant tissue plasminogen activator for acute emispheric stroke: the European Cooperative Acute Stroke Study (ECASS). JAMA 274: 1017–1025
3. The NINDS rt-PA Stroke Study Group (1995) Tissue plasminogen activator for acute ischemic stroke. N Engl J Med 333: 1581–1587

References

1. Alberts MJ, Pessin MS, Goldstein LB, Wolpert SM, Adams HP, del Zoppo GJ, and the rt-PA/Acute Stroke Study Group (1995) Predictors of cerebral edema after acute ischemic stroke (abstract). Stroke 26, 1: 184
2. Bogousslavsky J, Regli F, Uske A, Maeder P (1991) Early spontaneous hematoma in cerebral infarct: is primary cerebral hemorrhage overdiagnosed ? Neurology 41: 837–840
3. Bozzao L, Angeloni U, Fantozzi LM, Pierallini A, Fieschi C (1991) The value of early CT hypodensity in predicting middle cerebral haemorrhagic infarction. Neuroradiology 33 [Suppl]: 42–44
4. Bozzao L, Bastianello S, Fantozzi LM, Angeloni U, Argentino C, Fieschi C (1992) Correlation of angiographic and sequential CT findings in patients with evolving cerebral infarction. Am J Neuroradiol 10: 1215–1222
5. Brott TG, Haley EC, Levy DE, Barsan W, Broderick J, Sheppard GL, Spilker J, Kongable GL, Massey S, Reed R, Marler JR (1992) Urgent therapy for stroke: part I. Pilot study of tissue plasminogen activator administered within 90 minutes. Stroke 23: 632–640
6. Candelise L, for the MAST-I collaborative group (1993) The MAST-I study. In: del Zoppo GJ, Mori E, Hacke W (eds) Thrombolytic therapy in acute ischemic stroke II. Springer, Berlin Heidelberg New York Tokyo, pp 86–94
7. Cerebral Embolism Study Group (1984) Immediate anticoagulation of embolic stroke: brain hemorrhage and management options. Stroke 15: 779–789
8. Donnan GA, Davis SM, Chambers BR, States PC, Hankey GY, Stewart-Wyune EG, Rosen D, Tuck RR, McNeil JJ (1993) Australian Streptokinase Trial (ASK). In: del Zoppo GJ, Mori E, Hacke W (eds) Thrombolytic therapy in acute ischemic stroke II. Springer Berlin Heidelberg New York Tokyo, pp 80–85
9. Faris AA, Hardin CA, Poser CM (1963) Pathogenesis of hemorrhagic infarction of the brain. Arch Neurol 9: 468–472
10. Fisher CM, Adams RD (1951) Observations on brain embolism with special reference to the mechanism of hemorrhagic infarction. J Neuropathol Exp Neurol 10: 92–93
11. Gruppo Italiano per lo studio della sopravvivenza nell' infarto miocardico (1990) GISSI 2: a factorial randomised trial of alteplase versus streptokinase and heparin versus no heparin among 12490 patients with acute myocardial infarction. Lancet 336: 65–71
12. Hain RF, Westhaysen PV, Swank RL (1952) Hemorrhagic cerebral infarction by arterial occlusion: an experimental study. J Neuropathol Exp Neurol 11: 34–43
13. Hakim AM, Ryder-Cooke A, Melanson D (1983) Sequential computerized tomographic appearance of strokes. Stroke 14: 893–897
14. Haley EC, Levy DE, Brott TG, Sheppard GL, Wong MCW, Kongable GL, Torner JC, Marler JR (1992) Urgent therapy for stroke: part II. Pilot study of tissue plasminogen activator administered 91–180 minutes from onset. Stroke 23: 651–645
15. Hart RG, Easton JD (1986) Hemorrhagic infarcts. Stroke 17: 586–589
16. Hennerici M, Hacke W, Hornig C, Zangemeister W (1991) Intravenous tissue plasminogen activator for the treatment of acute thromboembolic ischemia. Cerebrovasc Dis 1 [Suppl]: 124–128
17. Hommel M, Boissel JP, Cornu C, Boutitie F, Lees KR, Besson G, Leys D, Amarenco P, Bogaert M, for the MAST Study Group (1995) Termination of trial of streptokinase in severe acute ischaemic stroke (letter). Lancet 345: 57
18. Hornig CR, Dorndorf W, Agnoli A (1986) Hemorrhagic cerebral infarction- a prospective study. Stroke 17: 179–185
19. Horowitz SH, Zito JL, Donnarumma R, Patel M, Alvir J (1991) Computed tomographic-angiographic findings within the first five hours of cerebral infarction. Stroke 22: 1245–1253
20. ISIS-2 Collaborative Group (1988) Randomised trial of intravenous streptokinase, oral aspirin, both or neither among 17187 cases of acute myocardial infarction. Lancet 2: 349–360
21. Jones T, Morawetz R, Crowell R (1981) Thresholds of focal cerebral ischemia in awake monkeys. J Neurosurg 54: 773–782
22. Kaplan B, Brint S, Tanabe J, Jacewicz M, Wang X, Pulsinelli W (1991) Temporal threshold for neocortical infarction in rats subjected to reversible focal cerebral ischemia. Stroke 22: 1032–1039

23. Kaste M, Fieschi C, Hacke W, Lesaffre E, Verstraete M, Frohlich J (1993) The European Cooperative Acute Stroke Study (ECASS). In: Del Zoppo GJ, Mori E, Hacke W (eds) Thrombolytic therapy in acute ischemic stroke II. Springer Berlin Heidelberg New York Tokyo, pp 66–71

24. Koudstaal PJ, Stibbe J, Vermeulen M (1988) Fatal ischaemic brain oedema after early thrombolysis with tissue plasminogen activator in acute stroke. BMJ 297: 1571–1574

25. Lodder J (1984) CT-detected hemorrhagic infarction: relation with the size of the infarct and the presence of midline shift. Acta Neurol Scand 70: 329–335

26. Lodder J, Krijne-Kubat B, Broekman J (1986) Cerebral hemorrhagic infarction at autopsy: cardiac embolic cause and the relationship to the cause of death. Stroke 17: 626–629

27. Lyden PD, Zivin JA (1993) Hemorrhagic transformation after cerebral ischemia: mechanisms and incidence. Cerebrovasc Brain Metab Rev 5(1): 1–16

28. Meyer JS, Gilroy J, Barnhart MI, Johnson JF (1963) Therapeutic thrombolysis in cerebral thromboembolism. Neurology 13: 927–937

29. Meyer JS, Gilroy J, Barnhart MI, Johnson JF (1965) Therapeutic thrombolysis in cerebral thromboembolism: randomised evaluation of intravenous streptokinase. In: Millikan CH, Siekert RG, Whisnant JP (eds) Cerebral vascular diseases. Fourth Princeton Conference. Grune and Stratton, New York, pp 200–213

30. Moulin T, Crépin-Leblond T, Chopard JL, Bogousslavsky J (1993) Hemorrhagic infarcts. Eur Neurol 34: 64–77

31. Okada Y, Yamaguchi T, Minematsu K, Miyashita T, Sawada T, Sadoshima S, Fujishima M, Omae T (1989) Hemorrhagic transformation in cerebral embolism. Stroke 20(5): 598–603

32. Ott BR, Zamani A, Kleefield J, Funkenstein HH (1986) The clinical spectrum of hemorrhagic infarction. Stroke 17: 630–637

33. Pessin MS, Teal PA, Caplan LR (1991) Hemorrhagic infarction: guilt by association ? Am J Neuroradiol 12: 1123–1126

34. Sette G, Toni D, De Michele M, Fiorelli M, Sacchetti ML, Cavalletti C, Gori C, Argentino C, Fieschi C (1994) Pharmacotherapy of stroke: an overview. In: Krieglstein J, Oberpichler-Schwenk H (eds) Pharmacology of cerebral ischemia 1994. Wissenschaftliche Verlagsgesellschaft, Stuttgart, pp 605–610

35. Sussman BJ, Fitch TSP (1958) Thrombolysis with fibrinolysin in cerebral arterial occlusion. JAMA 167:1705–1709

36. von Kummer R, Meyding-Lamadé U, Forsting M, Rosin L, Rieke K, Hacke W, Sartor K (1994) Sensitivity and prognostic value of early CT occlusion of the middle cerebral artery trunk. Am J Neuroradiol 15: 9–15

37. Wardlaw JM, Warlow CP (1992) Thrombolysis in acute ischemic stroke: does it work? Stroke 23(12): 1826–1839

38. Wardlaw JM, Lindley R, Warlow CP, Dennis MS, Sellar RJ (1992) Swelling in acute ischemic stroke and reperfusion: a prospective study in patients with large infarcts (abstract). Cerebrovasc Dis 2: 197

39. Yusuf S, Collins R, Peto R, Furberg C, Stampfer MJ, Goldhaber SZ, Hennekens CH (1985) Intravenous and intracoronary fibrinolytic therapy in acute myocardial infarction: overview of results on mortality, reinfarction and side-effects from 33 randomised controlled trials. Eur Heart J 6: 556–585

Correspondence: Cesare Fieschi, M.D., Department of Neurological Sciences, University of Rome "La Sapienza", Viale dell' Università 30, I-00185 Rome, Italy.

Acta Neurochir (1996) [Suppl] 66: 81–86
© Springer-Verlag 1996

Traumatically Induced Axonal Damage: Evidence for Enduring Changes in Axolemmal Permeability with Associated Cytoskeletal Change

J. T. Povlishock and **E. H. Pettus**

Department of Anatomy, Medical College of Virginia, Virginia Commonwealth University, Richmond, VA, U.S.A.

Summary

Recent studies have demonstrated that delayed or secondary axotomy is a consistent feature of traumatic brain injury in both animals and man. Moreover, these studies have shown that the pathogenesis of this secondary axotomy involves various forms of initiating pathology, with the suggestion that, in some cases, only the axonal cytoskeleton is perturbed, while, in other cases, both the axonal cytoskeleton and related axolemma manifest traumatically induced perturbations. In the current communication, we continue in our investigation of the significance of these traumatically induced alterations in axolemmal permeability and their relation to any related intra-axonal cytoskeletal change. This was accomplished in cats which received intrathecal infusions of peroxidase, an agent normally excluded by the intact axolemma. These animals were subjected to traumatic brain injury, and sites showing altered axolemmal permeability to the peroxidase were assessed at the light and electron microscopic level. Through this approach, we recognized that a traumatic episode of moderate severity evoked changes in axolemmal permeability which surprisingly endured for up to 5 hrs postinjury. At such focal sites of altered permeability, the related cytoskeleton showed a statistically significant neurofilament compaction, with the strong suggestion of concomitant neurofilament sidearm loss, microtubular dispersion, and mitochondrial abnormality. Over time, these events led to further disorganization of the axonal cytoskeleton which translated into impaired axoplasmic transport and secondary axotomy. Most likely, these alterations in axolemmal permeability result in either the direct or indirect effects upon the axonal cytoskeleton that precipitate the damaging sequences resulting in delayed axotomy.

Keywords: Traumatic brain injury; axonal damage; axolemma; cytoskeleton.

Introduction

Diffuse axonal injury has long been recognized as a feature of human traumatic brain injury that has been associated with morbidity and mortality [1,12,13]. While historically it was assumed that the tensile forces of injury tore axons at the moment of injury, work emerging from the experimental and clinical settings suggest that, with the exception of the most severe forms of traumatic brain injury [4], direct disruption of the axon cylinder does not occur [2,4,8,9,11,16]. Rather, it appears that a process of secondary or delayed axotomy predominates, whereby the traumatic episode elicits an impairment of axoplasmic transport that results in axonal swelling, with subsequent disconnection over a several-hour posttraumatic course [2,4,8,9,11,16]. While most investigators concur that the process of delayed or secondary axotomy is the predominant form of the axonal damage occurring with traumatic brain injury, new evidence suggests that the pathogenesis of traumatically induced delayed axotomy is quite complex, involving different forms of initiating pathobiology [6]. Specifically, with mild traumatic brain injury, injured axons appear to manifest only cytoskeletal change, whereas with more severe injury, the injured axons demonstrate cytoskeletal abnormalities whose genesis appears to be influenced by concomitant changes in axolemmal permeability [6]. Specifically, in those axons showing altered axolemmal permeability, the neurofilaments appear to undergo rapid compaction, followed over time by a relatively delayed misalignment, leading to impaired axoplasmic transport and delayed axotomy [6]. This is in contrast to those axons showing primary cytoskeletal change which manifest rapid neurofilament misalignment without compaction as a prelude to the disruption of axoplasmic transport and delayed axotomy.

In the current communication, we continue in our analysis of those abnormalities associated with traumatically induced alterations in axolemmal permeability. In this context, we seek to define fully the posttraumatic duration of this traumatically induced alteration in axolemmal permeability, while fully explicating, through quantitative methods, the precise cytoskeletal/neurofilament changes that occur in relation to these sites of altered axolemmal permeability.

Materials and Methods

Essentially, the protocols used in the current investigation closely adhere to those previously reported [6]. Sixteen adult male cats were anesthetized with sodium pentobarbital (30 mm/kg iv), and were surgically prepared for the induction of a fluid-percussion traumatic brain injury consistent with protocols described in previous communications [14]. The animals evaluated in the current investigation were subjected to injuries of moderate to moderately severe intensity, ranging from 2 to 2.8 atms in severity, and were allowed to survive for varying periods ranging from 5 mins to 8 hrs postinjury. To examine the issue of traumatically induced alterations in axolemmal permeability, 50 mg of horseradish peroxidase (HRP) Type VI (Sigma Chemical Co., St. Louis, MO) was dissolved in autologous CSF and infused into the cisterna magna via a 25 gauge spinal needle. Previous experience with this approach has shown that the intrathecally infused HRP diffuses into the extracellular space of brain stem parenchyma where, in the control situation, its intra-axonal passage is restricted by the presence of the intact axolemma [6]. In all animals, the intrathecal peroxidase was allowed to remain *in situ* for 1–2 hrs, prior to the induction of the traumatic fluid-percussion insult. Thus, in those animals with a relatively short posttraumatic survival time, the intrathecal infusion was accomplished prior to the induction of the traumatic brain injury. Alternatively, for those animals surviving for a more prolonged posttraumatic course, the intrathecal infusion was initiated 1–2 hrs prior to sacrifice.

At the designated survival times, the animals received an overdose of sodium pentobarbital (100 mg/kg), and were transcardially perfused with aldehydes. The brain stems were removed the bisected, with one bisected segment undergoing sagittal sectioning at 50μm on a vibratome. The other bisected segment was coronally sectioned in the same fashion. All sections were then reacted for the visualization of HRP reaction product, utilizing 0.05 % diaminobenzidine, 0.2 % B-D glucose, 0.04 % ammonium chloride and 0.00041 % glucose oxidase in a 0.1M sodium phosphate buffer. Alternate sections were then processed for either light- or electron microscopy. In this approach, the sagittally harvested sections were used to ascertain, if the traumatic insult was capable of eliciting alteration in axolemmal permeability to the extracellurly confined HRP, while also determining, if any observed alteration in axolemmal permeability was an enduring posttraumatic event.

The sagittal sections were also used to identify those anatomical foci, revealing peroxidase-containing axons so that comparable foci could be selected in the coronal sections for subsequent detailed analysis of intra-axonal cytoskeletal change. To this end, coronally sectioned axons demonstrating intra-axonal HRP were photographed at a magnification of 50,000X, and these were compared and contrasted to size-matched axons demonstrating no evidence of HRP uptake. Both HRP and non-HRP-containing axons were then overlaid by a standardized grid containing hexagons, each encompassing 0.025 μm (Fig. 2). These micrographs were then transferred to an image analysis unit where they were videographically captured, digitized and enlarged. In each grid, the neurofilament density/number was computed, with a total of 20 grids evaluated per micrograph. In all, 5 HRP-containing and 5 non-HRP-containing axons were analyzed in this fashion, and statistical evaluation of these findings was performed.

The controls used in the current investigation varied. As noted above, non-HRP-containing axons served as internal controls, allowing for statistical comparisons to be made between the HRP-containing and non-HRP-containing groups. In addition to these controls, two (2) other animals received intrathecal peroxidase but were not subjected to injury. These animals were subsequently prepared and processed in accordance with the above described protocols.

Results

Animals subjected to traumatic brain injury showed evidence of focal intra-axonal peroxidase flooding which was most prominent in the pontomedullary and cervicomedullary junctions (Fig. 1). As previously suggested, such peroxidase flooding was seen within minutes of the traumatic episode, clearly indicating that alterations in the axolemma's permeability to the extracellularly confined peroxidase occurred early in the posttraumatic course. Not only was such altered axolemmal permeability to peroxidase seen within the first minutes postinjury, but also, it continued over a relatively long posttraumatic course. Such alterations in the axolemma's permeability persisted for at least 5 hrs postinjury, the point at which the HRP was first infused in those animals surviving for up to 7 hrs. However, in one animal, when the HRP was infused at 7 hrs postinjury with an additional hour of posttraumatic survival, no intra-axonal HRP was observed, suggesting a restitution of the axolemma's traumatically altered permeability at this relatively late posttraumatic time point.

When the HRP-containing axons were visualized in the coronal plane and compared and contrasted to size-matched non-HRP-containing axons, a distinct

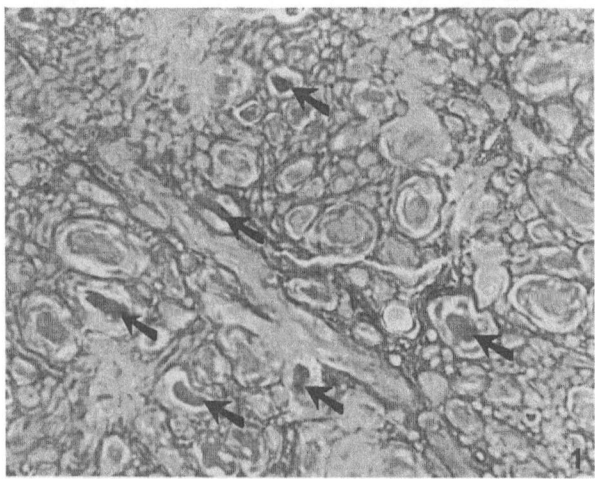

Fig. 1. Light micrograph of a plastic thick section of a brain stem cut in the coronal plane. Note the HRP-containing axons (arrows), some of which appear dense and somewhat irregular, suggesting shrinkage from the overlying myelin sheath. (X 1,000)

set of neurofilament and cytoskeletal changes were recognized. Within 5 mins of the traumatic episode, a dramatic increase in the number of neurofilaments per unit area occurred in the HRP-containing axons (Figs. 2–4). In those axons containing HRP, the mean neurofilament density/number per 0.025 μm² reached 37.97 ± 5.93 (Mean ± SEM) in contrast to the non-HRP-containing axons in which the neurofilament density reached only 18.67 ± 2.88 (Mean ± SEM) per 0.025 μm². These differences were highly significant ($t_4 = 3.98$, $p<0.05$). Despite their increase per unit area which reflected increased NF compaction, the neurofilaments did not change in their alignment, maintaining their normal linear course parallel to the axon's long axis. Careful examination of the coronal sections from HRP and non-HRP-containing axons revealed other differences in addition to this statistically significant increase in neurofilament density. Specifically, in the HRP-containing an apparent loss of the neurofilament sidearms occurred, and this loss was striking in comparison to the non-peroxidase-containing as well as those axons harvested from the sham controls (Figs. 2 and 3). Further, although not quantitatively validated, there was a strong suggestion of microtubular loss as well as local damage to the mitochondria, reflected in mitochondrial swelling with disruption of their cristae. These intra-axonal events were also associated with a marked irregularity and infolding of the axon cylinder, with the suggestion that the axon was shrinking away from its overlying myelin investment. Typically, all of these events were seen as early as 5 mins postinjury, and these changes were observed in both nodal and inter-nodal regions. When these coronally sectioned axons were followed over time, it was somewhat surprising to find that the above described intra-axonal changes persisted relatively unchanged for a period of up to 2–3 hrs postinjury. In these cases, the axons displayed the typical repertoire of increased filament density associated with sidearm loss, microtubular dispersion, and mitochondrial damage. With continuing survival, the same axons manifested continued evidence of neurofilament change, this time reflected in an alteration in neurofilament alignment. Now, the neurofilaments moved in planes oblique to the long axis of the axon, and at these sites, this altered neurofilament alignment was associated with an accumulation of organelles, most likely the result of impaired axoplasmic transport.

Discussion

The results of the current investigation confirm and significantly extend our previous observations regarding the occurrence of traumatically induced change in axolemmal permeability and its relation to ongoing alterations in the axonal cytoskeleton [6]. First, the current communication confirms that moderate traumatic brain injury can evoke altered axolemmal permeability change. It significantly extends previous observations by showing that such permeability change can persist for up to 5 h postinjury, while suggesting that beyond this time, there is a restitution of the axolemma's ability to exclude HRP passage. The ultrastructural correlates of this altered axolemmal permeability are, at the moment, unknown. However, in the studies conducted to date, no evidence of direct membrane renting has been detected. Conceivably, the traumatic episode causes subtle membrane perturbations not detectable by routine ultrastructural analyses.

Although the finding of traumatically induced alterations in axolemmal permeability is of interest, it becomes even more significant when one considers its biological implications. Normally, the intact axolemma helps to maintain the intra-axonal environment which varies considerably from that found in the extracellular compartment. Specifically, with disruption of the axolemma of the magnitude observed, one could easily conceptualize, how various excluded ions could move from the extracellular space to reach the intracellular compartment where they could exert either direct or indirect effects upon the axonal cytoskeleton.

As noted in the current communication, those axonal sites demonstrating altered axolemmal permeability also manifest concomitant, rapid cytoskeletal change reflected in a statistically significant increase in neurofilament density. This increase in density appears to correlate with the loss and/or collapse of the neurofilament sidearms which, in turn, allows for increased neurofilament compaction, translating into a general collapse/shrinkage of the axon cylinder. Interestingly, this loss or collapse of the neurofilament sidearms, with concomitant neurofilament compaction, also appears to be associated with microtubular loss and mitochondrial abnormalities, although these findings have not been statistically validated.

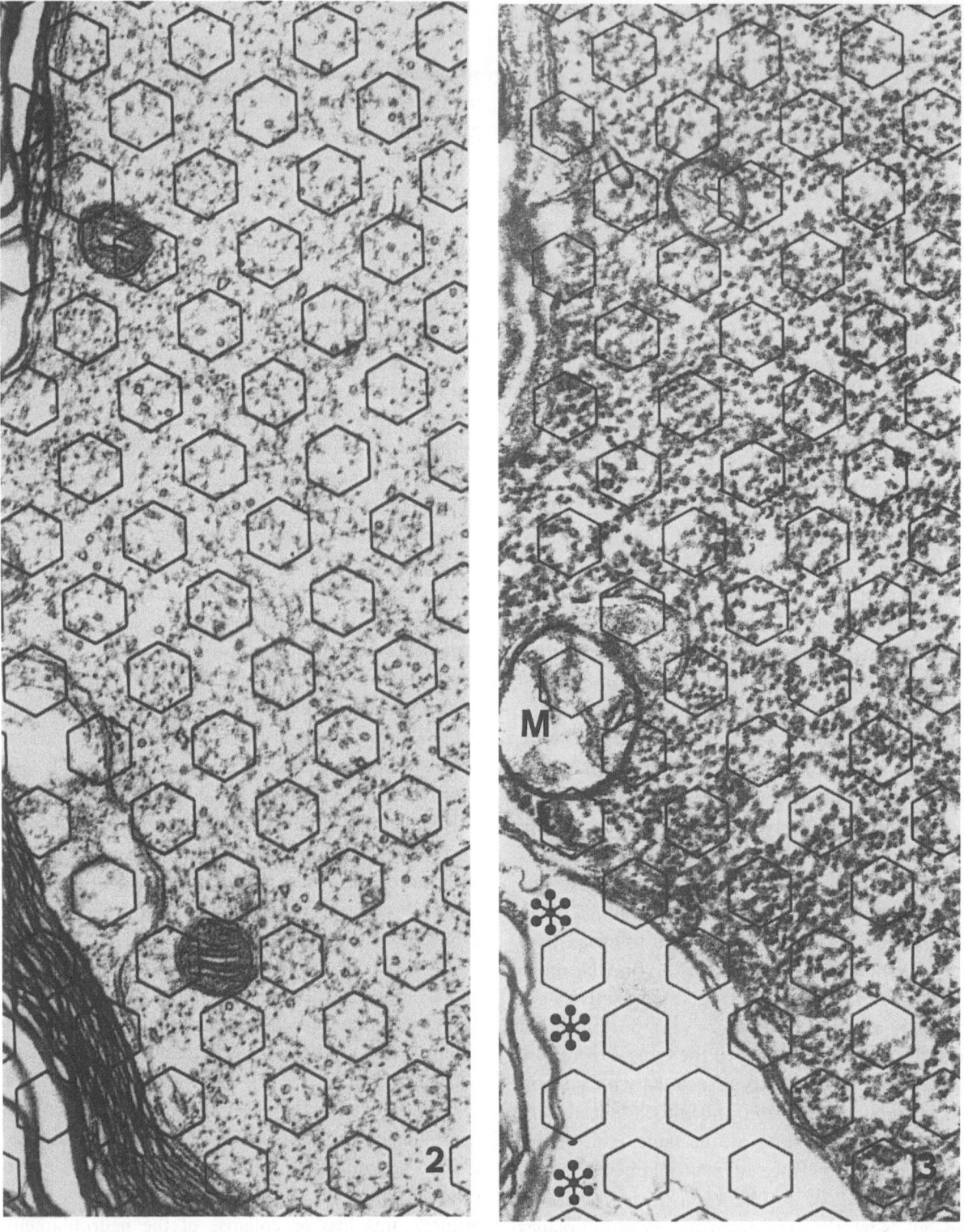

Figs. 2 and 3. Ultrastructural details of a non-HRP-containing (Fig. 2) and a HRP-containing axon (Fig. 3). The hexagonal grid which overlies both axons was used in the quantitative assessment of the filament density. Note that the neurofilaments seen in the non-HRP-containing axon (Fig. 2) appear widely dispersed, with conspicuous sidearms projecting between the transversely sectioned neurofilaments. Also, note that these neurofilaments are associated with unaltered mitochondria. In contrast to this, note that the neurofilaments seen in the HRP-containing axon (Fig. 3) appear densely packed with an apparent loss of their sidearms. Also note that these neurofilament abnormalities are associated with local mitochondrial damage (*M*) as well as shrinkage of the axon away from the overlying myelin sheath (*). It should be noted that although this HRP-containing axon demonstrates traumatically induced neurofilament packing, other HRP-containing axons examined in this study showed even more dramatic packing. (x 50,000)

Frequency Distribution

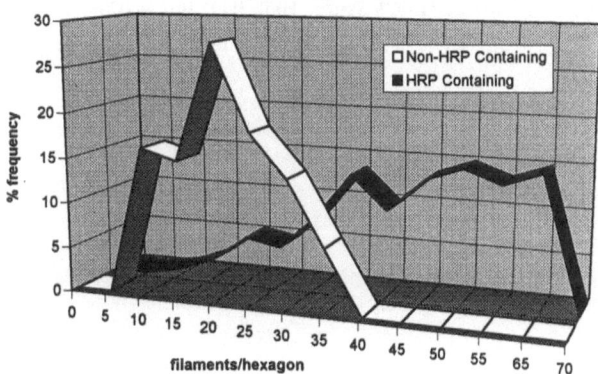

Fig. 4. Frequency distribution of the neurofilament numbers/densities in HRP-containing vs non-HRP-containing axons. Note that in the HRP-containing axons, there is a conspicuous increase in neurofilament number/density

While collectively these events seem somewhat consistent with an intra-axonal dysregulation in calcium homeostasis with an activation of neutral proteases [7,10,15], it is somewhat surprising that these events occurred without evidence of any continued neurofilament degradation, since such dissolution has been posited to occur subsequent to a disruption of calcium homeostasis [10]. In this context, however, it is conceivable that the calcium may depolymerize microtubules and activate proteases that cleave the neurofilament sidearms, resulting in altered neurofilament structures that are resistant to further enzymatic degradation. Alternatively, it is possible that these events may not be related to local calcium dysregulation. Rather, they may be related to other ionic or secondary messenger changes. In this context, preliminary work conducted in our laboratory, using the calcium chelator BAPTA-AM, has failed to demonstrate any significant protection in terms of the observed neurofilament alterations. Since it is well-known that the neurofilament sidearms are highly phosphorylated, it is possible that other events may trigger dephosphorylation, resulting in their apparent collapse [5]. Obviously, these issues require continued investigation in order to dissect out causality and thereby determine if any rational therapeutic interventions can be developed.

In addition to the observations made during the acute phase of peroxidase influx, the results of the current investigation are equally important in that they confirm that once initiated, the increased neurofilament packing remains relatively stable for a several hour course after which time the neurofilaments begin to disassemble, together with the onset of impaired axoplasmic transport. The considerable delay involved

in these evolution events is somewhat surprising and suggests that once initiated the neurofilament compaction may set into motion a series of secondary changes that result in ultimate neurofilament disassembly/disruption. Precisely how this is accomplished is unknown, but perhaps secondary messenger systems evoked by the traumatic insult lead to this delayed neurofilament disassembly [5].

Acknowledgements

The authors gratefully acknowledge the skilled technical assistance of Susan Walker and Mary Lee Giebel. Supported by NIH Grant NS20193.

References

1. Adams JH, Graham DI, Murray LS, Scott G (1982) Diffuse axonal injury due to nonmissile head injury in humans: An analysis of 45 cases. Ann Neurol 12: 557–563
2. Erb DE, Povlishock JT (1988) Axonal damage in severe traumatic brain injury: an experimental study in cat. Acta Neuropathol 76: 347–358
3. Grady MS, McLaughlin MR, Christman CW, Valadka AB, Fligner CL, Povlishock JT (1993) The use of antibodies targeted against the neurofilament subunits for the detection of diffuse axonal injury in humans. J Neuropath Exp Neurol 52: 143–152
4. Maxwell WL, Irvine A, Graham DI, Adams JH, Gennarelli TA, Tipperman R, Sturatis M (1991) Focal axonal injury: the early axonal response to stretch. J Neurocytol 20(3): 157–164
5. Nixon RA, Sihag RK (1991) Neurofilament phosphorylation: a new look at regulation and function. TINS 14 (11): 501–505
6. Pettus EH, Christman CW, Giebel ML, Povlishock JT (1994) Traumatically induced altered membrane permeability: its relationship to traumatically induced reactive axonal change. J Neurotrauma 11: 507–522
7. Pierluigi N, Orrenius S (1992) Ca^{2+} and cell death. Ann N Y Acad Sci 648: 17–27
8. Povlishock JT (1992) Traumatically induced axonal injury: pathogenesis and pathobiological implications. Brain Pathol 2: 1–12
9. Povlishock JT, Becker DP, Cheng CLY, Vaughan GW (1983) Axonal change in minor head injury. J Neuropath and Exp Neurol 42: 225–242
10. Schlaepfer WW (1987) Neurofilaments: structure, metabolism and implications in disease. J Neuropath Exp Neurol 46: 117–129
11. Sheriff FE, Bridges LR, Sivaloganathan S (1994) Early detection of axonal injury after human head trauma using immunocytochemistry for β-amyloid precursor protein. Acta Neuropathol 87: 55–62
12. Strich SJ (1956) Diffuse degeneration of the cerebral white matter in severe dementia following head injury. J Neurol Neurosurg Psychiatry 19: 163–185
13. Strich SJ (1970) Lesions in the cerebral hemisphere after blunt head injury. J Clin Pathol 4: 166–171
14. Sullivan HG, Martinez AJ, Becker DP, Miller JD, Griffith R, Wist AO (1976) Fluid-percussion model of mechanical brain injury in the cat. J Neurosurg 45: 520–534
15. Waxman SG, Black JA, Ransom BR, Stys PK (1993) Protection of the axonal cytoskeleton in anoxic optic nerve by decreased extracellular calcium. Brain Res 614: 137–145

16. Yaghmai A, Povlishock J (1992) Traumatically induced reactive change as visualized throughout the use of monoclonal antibodies targeted to neurofilament subunits. J Neuropath Exp Neurol 51(2): 158–176

Correspondence: John T. Povlishock, Ph.D., Department of Anatomy, Room 11–071 Sanger Hall, P.O. Box 980709, 1101 E. Marshall Street, Richmond, VA 23298–0709, U.S.A.

Acta Neurochir (1996) [Suppl] 66: 87–95
© Springer-Verlag 1996

Expression of Microglial Markers in the Human CNS After Closed Head Injury

S. Engel[1], **H. D. Wehner**[2], and **R. Meyermann**[1]

[1]Institute of Brain Research, and [2]Institute of Forensic Medicine, University of Tübingen, Tübingen, Federal Republic of Germany

Summary

The loss of neurons after severe closed head injury is not only a consequence of the primary impact but also of secondary damage mechanisms. Among the cell population of the central nervous system microglia is surely a candidate to influence secondary damage mechanisms by releasing cytotoxic cytokines [1]. About microglial reaction in closed traumatic brain injury (TBI), however, no data are available. In contrast to experiments using stab wound injury the coverings of the CNS in closed TBI are still intact.

We have examined 17 patients who died because of TBI after various times post injury. We studied the expression of antigens which are either permanently present on microglial cells or those which are only facultatively found on activated microglia.

Low numbers of microglial cells were shown to express MHC-class II antigens immediately after TBI which is also true for CD 68 and leukocyte common antigen (LCA). Surprisingly, however, antigens such as HAM 56 were expressed not earlier than 72 hours after TBI as well as the lectin *ricinus communis* agglutinin-1 (RCA-1). The results indicate a delayed activation of microglia in traumatic brain injury.

Keywords: Head injury; microglia; macrophages.

Introduction

The role of microglial activation following pathological events has been of interest since their first descriptions by Nissl (1894) and Rio-Hortega (1932) [3,19]. Experimental models to study the mechanisms of microglial activation require an approach which does not interrupt the integrity of the central nervous system (CNS) because activated microglia cannot be distinguished from blood born macrophages. This approach was utilized in the so called "facialis model" (for rev. see Kreutzberg, this volume pp. 103). Cutting the peripheral root of the facialis nerve results in proliferation of microglial cells and their migration to a perineuronal position. In contrast, additional application of neurotoxic drugs induce neurotoxic mechanisms of microglial cells. These different functions of microglia represent either different phases or different pathways of activation.

Traumatic brain injury (TBI) is one of the rare human diseases in which the time of onset of the pathological changes is exactly known. The first pathological changes are characterized by a local edema within minutes, perivascular hemorrhages and the presence of polymorphonuclear cells within hours. Degeneration of neurons and axonal swelling as well as ballooning and disconnection can be detected in each case caused by either mechanical forces or secondary damage mechanisms. Macrophages invade the contusion area 2 hours post TBI and might change into siderophages after about 72 hours. Lymphocytes can be detected 3–4 days post injury. The primary lesions are followed by many additional secondary lesions depending on severeness and kind of injury for example circulation disturbances, local and general ischemia, inflammation, edema and swelling of parenchyma leading to herniation of the brain. (for rev. see [26]).

A candidate which might influence secondary damage of the tissue is the microglial cell. This cell is equipped with a cascade of mechanisms which can both, enhance and downregulate tissue damage. Due to this idea we have studied microglial reaction in TBI. Knowledge about microglial reaction offers different advantages: (1) Better understanding of pathomechanisms and therapeutic strategies, (2) Detailed typing and staging of lesions which is important in forensic medicine.

This problem was approached by immunocytochemical methods. The microglia was labeled by markers which are either constitutively or facultatively expressed on the microglial cell population but which are also found on macrophages and monocytes.

Table 1. *Clinical and Autoptical Data*

No.	Age	Sex	Time of survival	Kind of injury	Edema, herniation, coma			SDH, SAB	Cause of death	Clinical data
1	27	w	hours	traffic accident	+	+	+	SDH	brain death	trep.
2	36	m	hours	brawl and fall	+	+	+	SAB	brain death	
3	43	m	hours	fall	+	–	ni	–	brain death	chronic alcoholism
4	26	w	12 hours	traffic accident	+	–	+	–	hemorrhagic shock	trep.
5	28	m	14 hours	traffic accident	+	+	+	SAB, SDH	hemorrhagic shock	DAI
6	52	m	12–24 hours	brawl and fall	+	+	+	SDH	brain death	gliding contusion
7	48	w	36 hours	brawl and fall	+	+	+	SDH	brain death	DAI
8	79	m	72 hours	brawl and fall	+	+	+	SDH	brain death	trep., older ischemia
9	30	w	72–96 hours	fall	+	+	+	–	brain death	pregnancy, sinus cavernosus fistula
10	55	m	8 days	traffic accident	+	+	+	–	brain death	skfrac.
11	41	m	10 days	traffic accident	+	–	ni	–	brain death	brain stem contusion, skfrac., DAI
12	87	m	14 days	fall	+	–	ni	SAB, SDH	brain death	hemorrhagic insult 48–72 h, skfrac.
13	83	m	14 days	traffic accident	+	+	+	SDH	brain death	intracerebral hematoma 48–72 h, trep.
14	80	w	15 days	traffic accident	+	+	+	–	brain death	skfrac.
15	18	w	16 days	traffic accident	+	+	+	–	multi organ failure	trep., reanimation, anoxia
16	53	m	21 days	brawl and fall	+	–	–	EDH	pneumonia	ischemic injury 4 days, skfrac., trep.
17	56	w	6 months	traffic accident	+	+	+	SDH	pneumonia	skfrac., hemiplegia, ischemic injury

SAB subarachnoidal bleeding, *SDH* subdural hematoma, *EDH* epidural hematoma, *DAI* diffuse axonal injury, *skfrac.* skull fractures, *trep.* trepanation, *ni* no information.

Material and Methods

Macroscopy and Histology

17 brains from deceased patients with severe TBI were autopsied. The selected brains had a distinct survival time from several hours to six months post TBI. The clinical and autoptical findings are summarized in Table 1. Samples of the formalin fixed brains were taken from 1.) the contusional area, 2.) tissue of the immediate vicinity, 3.) from non-injured areas and 4.) the brainstem.

The formalin-fixed and paraffin-embedded tissue was sectioned, routinely processed and stained with hematoxylin-eosin, Elastica-van-Gieson, Carstairs, Bielschowky's axon staining, Luxol-fast-blue and Berlin-Blue to determine the borders of the tissue injury. Other pathological events than those due to the mechanical trauma were excluded. Only in one case, number 3 in Table 1, we got knowledge of alcoholism, which had induced probably pretraumatic microglial reaction.

Immunohistochemistry

The following antibodies reactive with antigens expressed by either resting or activated microglia were used: Anti- CD 68 (clone KP-1, DAKOPATTS, Hamburg, FRG) [13], anti- human leukocyte common antigen (LCA, CD 45, clone PD 7/26 and 2B11, DAKOPATTS, Hamburg, FRG) (14), anti- monocytes/tissue histiocytes (MAC 387, DAKOPATTS, Hamburg, FRG), [13] anti-human alveolar macrophage (HAM 56, ENZO Diagnostics, New York NY, USA) [13]. Microglia was also studied with the biotinylated lectin *ricinus communis* agglutinin-1 (RCA-1, Vector Burlingame CA, USA)

[2,13,17]. For detection of MHC-II antigens, we used HLA-DR (clone CR3/43, DAKOPATTS, Hamburg, FRG) [7,13,18,22]. For demonstration of astrocytic response, a monoclonal antibody against glial fibrillary acid protein (GFAP, Boehringer, Mannheim, FRG, clone G-A-5) was applied. Axon spheroids were visualized by both anti-Neurofilament 68 kD light chain antibodies (clone NR 4, Boehringer, Mannheim, FRG) and anti-Alzheimer precursor protein A4, (βAPP-A4, clone 22C11, Boehringer, Mannheim, FRG) [10,11]. Antigens were exposed either by trypsin (CD 68, LCA) or by microwave pretreatment in citrate-buffer (pH 6.0) (NF 68-kD, βAPP-A4). Endogenous peroxidase was blocked with 0.3 % hydrogen peroxide in 100 % methanol. Slices were incubated with 10 % Tris- buffered (pH 7.1) normal swine serum containing 0.1 % bovine serum albumin (BSA) to prevent unspecific binding. The specific binding of the primary antibodies was followed by biotinylated anti-mouse or anti-rabbit F (ab') 2 fragments (DAKOPATTS, Hamburg, FRG). The avidin-biotin peroxidase complex technique (DAKOPATTS, Hamburg, FRG) with 3,3'-diaminobenzidine (DAB) as substrate was used to visualize the specific antigen binding. All sections except those labeled with NF-68 and βAPP were counterstained with hematoxylin. Specificity of the method was controlled by omission of the primary antibody.

Light Microscopy and Counting

Positively labeled structures which had the appearance of either ramified or ameboid microglia with a visible nucleus were counted light microscopically in a 50 mm^2 grid of each selected area. Cells which could not be distinguished from phagocytosing macrophages were not included. Grey and white matter was evaluated separately.

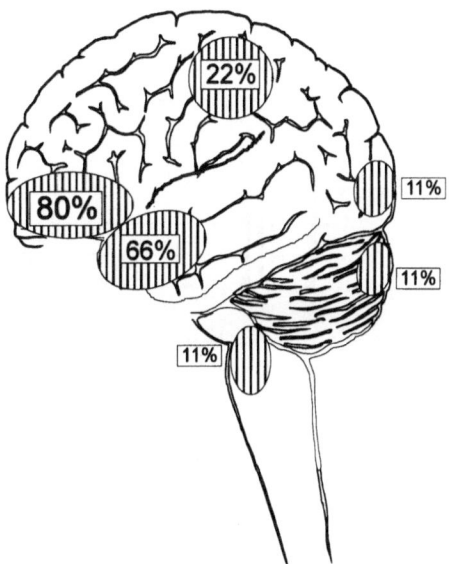

Fig. 1. The frequency of injury was up to 80 % in the frontal lobe, whereas the parietal (66 %) and temporal lobe (22 %) were less frequently injured. The lowest frequency was found in the cerebellum, the brain stem and the occipital lobe. n=17

Statistical Methods

The Wilcoxon Kruskal-Wallis tests (signed rank) was used to correlate the group with survival of less than 72 hours post TBI with the group with survival of more than 72 hours post TBI. To compare differences between grey and white matter we subtracted the counts in white from those in grey matter based on the Students t-test. The nonparametric measures of association were used to examine the association of markers.

Results

General Findings and Routine Histology

Most of the TBI lesions were found in the frontal lobe, whereas cerebellum and brain stem were affected in only 11 % of the brains studied. The localizations of contusional areas are listed in Fig. 1. Only in two cases a contrecoup opposite to the injured area could be identified. The histological evaluation revealed pathological changes which corresponded exactly with the survival time after the accident.

In all of the 17 cases microglia could be labeled with the described markers, except for MAC-387 antibodies which did not label any microglial cells. For control, adjacent sections were stained by anti-GFAP antibody to ensure the proper immunocytochemical reaction conditions.

Axon Spheroids Detected with βAPP-A4 and NF-68

We have visualized axon spheroids or swollen axons with antibodies against NF-68 (14 of 17) and β-APP-

A4 (14 of 17). Only one case lacked such pathological changes of axons. Two other cases showed axon spheroids labeled only by one of the antibodies. Axon spheroids were found even in cases with only a few hours of survival time. A high number of axon spheroids was detected in the region adjacent to the injured area. Numerous spheroids, however, were labeled in the brain stem, indicating that these patients suffered from additional brain stem contusion or diffuse axonal injury (DAI).

Microglial Marker Expression up to 72 hours

As shown by staining with antibodies against HLA-DR, CD 68 and LCA in normal brains, in injured brains microglia was also more frequent in the white than in the grey matter (Fig. 2, A–I and Fig. 3, A,C). This was proven for all tested regions (signed rank t-test between Prob>t: 0.016 and 0.031). The association of these three markers corresponded significantly: (HLA-DR-CD 68: Prob</Rho/ 0.0006; LCA-CD 68 Prob</Rho/ 0.0231). Labeling with anti-CD 68 cells showed a stronger staining than with antibodies against HLA-DR and LCA. The cells were rod shaped with small processes, occurring prominently in perivascular areas. Regarding the selected areas, the labeled microglia showed no striking differences. A slight increase in cell number was detected only in the vicinity of the trauma (see Fig. 2). The brain stem, however, showed many positively labeled cells without distinct distribution (not shown).

Before 72 hours post TBI, only endothelium and blood-born macrophages, which occasionally were found perivascularly and in the parenchyma, could be detected with RCA-1 and HAM 56. One case (number 3 in Table 1) was an exception which showed highly ramified RCA-1 positive cells in some distinct areas which could not be related to the trauma. This patient suffered from chronic alcoholism. A long slender cytoplasm with many branches classified these cells to the resting type of microglia. The number of cells was highly increased compared to other cases with a survival time of less than 72 hours (Fig. 4, A,C,G–I).

Microglial Reaction after more than 72 Hours Post TBI

72 hours post TBI microglial cells expressed HAM 56 and RCA-1 in both grey and white matter. Only in one case (see above) RCA-1 was expressed earlier. These findings were highly significant. (RCA-1 Prob>/Z/ 0.0018; Prob>ChiSq 0.0015; HAM 56Prob>/Z/

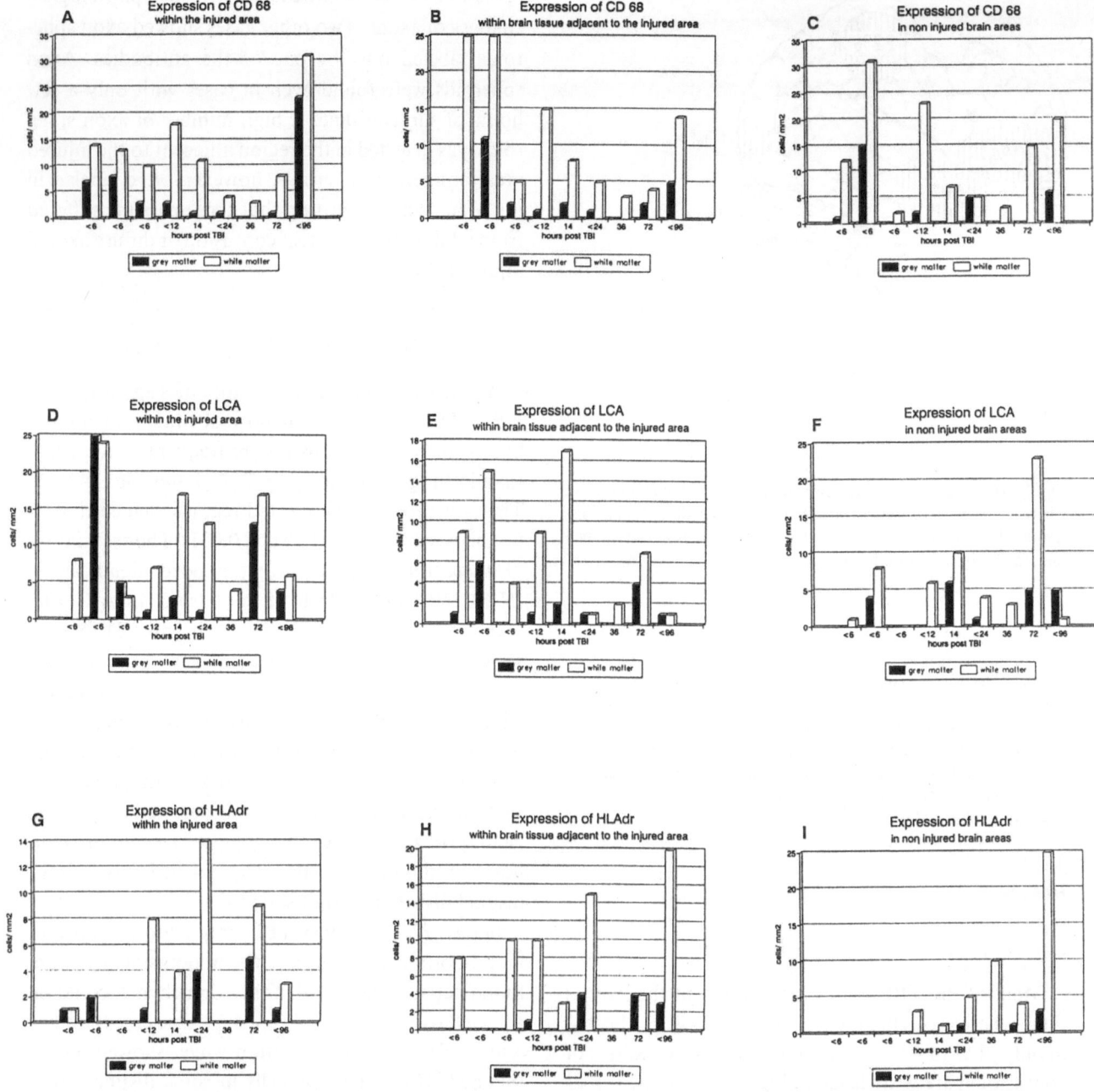

Fig. 2 A–I. In all examined regions microglia was labeled by antibodies against HLA-DR, LCA, and CD 68 from the beginning. Not all markers were labeling an identical number of cells. Microglia was more frequently found in the white matter than in the grey matter. Even in non- injured brain tissue microglia- labeling showed the same pattern, which was also true for the brain stem (not shown)

Fig. 3. Approximately 6 hours post TBI microglial cells did not express all antigens which are found on both: activated microglia and macrophages. Yet, CD 68 antigens were present from the beginning in activated microglial cells (A, arrows). The photomicrograph was take from the brain stem of patient number 1 (see table 1) who died few hours post TBI. The same area did not express HAM 56, an antigen isolated from alveolar macrophages (B). Microglia in the vicinity of a traumatic lesion did also express CD 68 antigens (C, arrows). In the same area the lectin RCA-1 was labeling only endothelial cells but not microglia (D). (bar = 25 µm). More than 72 hours post TBI labeling of microglia cells followed a quite different pattern. HAM 56-antigen was found then not only on endothelial but also on microglial cells (arrows), in a macroscopically not injured area, as shown in (E), demonstrating a case with 10 days survival. The same is true for labeling with RCA-1 as shown in (F) (arrows). RCA-1 labeled microglia was now also found at the site of the lesion (G) together with macrophage- like cells. MHC-II antigens, however, were only very weakly expressed (arrows in H). (bar = 40 µm)

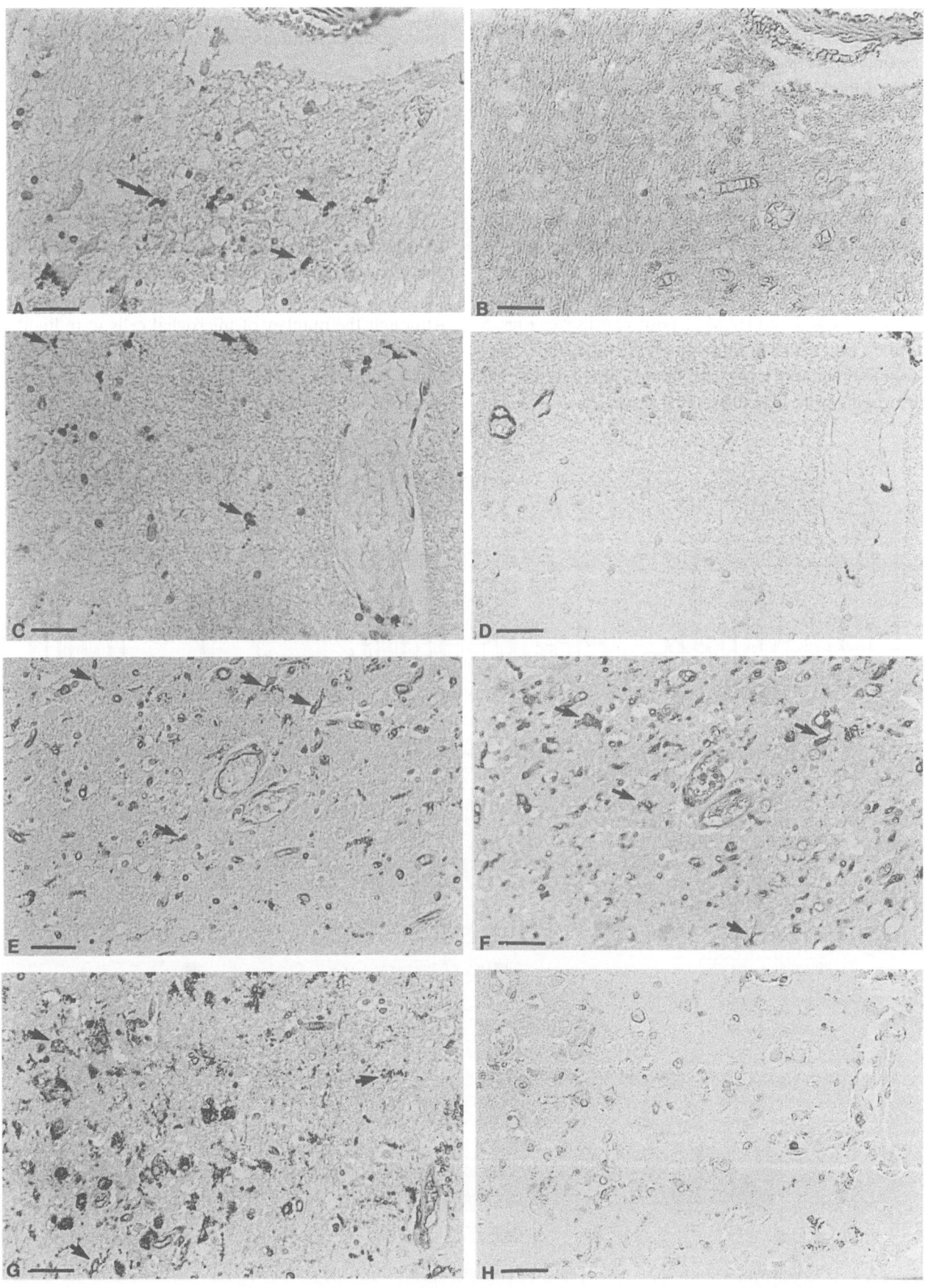

0.0004; Prob>ChiSq 0.0003) (Fig. 4). Calculating the association between CD 68, HLA-DR and LCA against HAM 56 and RCA-1, no significant differences were found. Based on the appearance and shape of cells these markers were not present on identical cells. Mainly in the area of contusion, where cell debris was found, RCA-1 and HAM 56 positive cells had large somata with short stubby processes, and lots of cells could not be distinguished from macrophages (Fig. 3, F, G). HLA-DR, CD 68 as well as LCA positive cells were more ramified and heterogenous in morphology. These cell were found in areas with no RCA-1 and HAM 56 positive cells (see Figs. 5 and 6), ranging from zero up to more than 40 cells per counted region (Fig. 4 A,B). In the single case labeled microglial cells were distributed equally in all selected areas. Summarizing the different staining patterns, all microglial cell numbers were increased after 72 hours post TBI.

Discussion

In traumatically injured human brain it is currently shown that up to 72 hours post traumatic brain injury (TBI) microglial cells express HLA-DR-antigens, LCA and the CD 68 molecules. The expression of these markers were found in the injured areas and nearby as well as in areas far away from the damaged tissue. Although the number of microglial cells were highly heterogenous [16], their number was significantly higher in the white than in the grey matter as described for the normal CNS [4,7,12,22]. These findings confirm

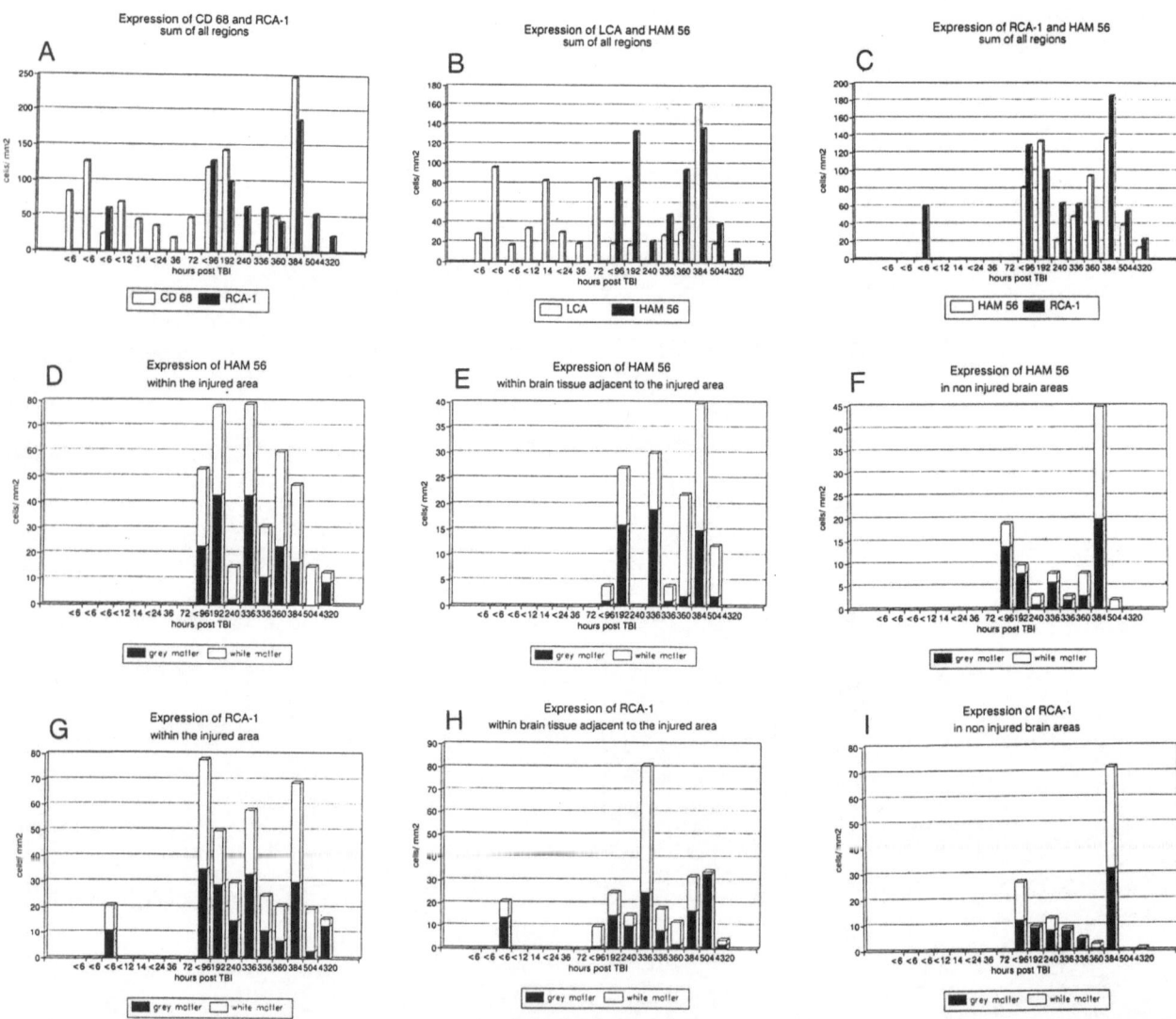

Fig. 4 A–I. LCA antibodies were labeling in most of our cases microglia within the observed time course. The same was true for anti- CD 68 (A,B), whereas RCA-1 and mAb HAM 56 were not found before 72 hours post TBI (A–I). This was not restricted to injured areas, but also found in the other areas studied

a permanent expression of HLA-DR molecules and LCA antigens [4,7,12,21,22].

In experimental studies (rev. by Kreutzberg this volume, [25]) a very early proliferation of microglial cells

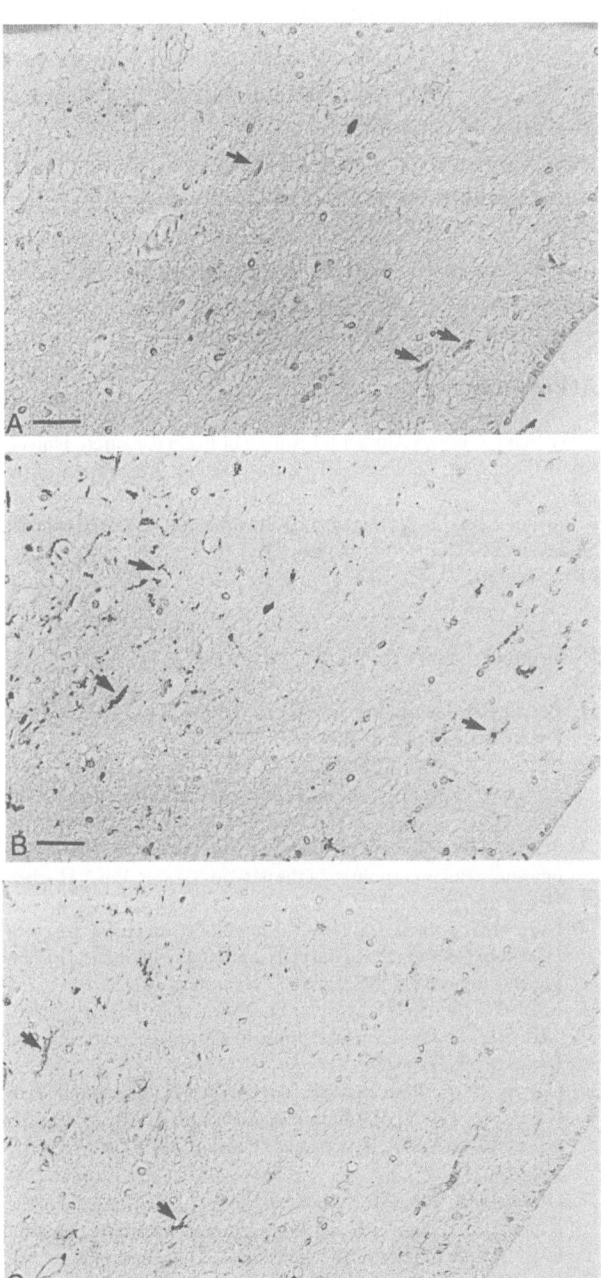

Fig. 5. Although all markers were observed on activated microglial cells, they were not equally expressed at a given time. In periventricular subependymal tissue far away from the impact (16 days post TBI), LCA was only weakly expressed on ramified microglia (A, arrows). CD 68, however, was also found in the deeper subependymal layers on rod shaped and more ramified cells (B, arrows). Distribution of MHC II- antigens had quite a similar pattern but occurred much less frequent than CD 68 (C, arrows). MHC-II was present only on ramified cell types in the deeper layers and adjacent to the ependyma. (bar = 40 μm)

has been described following cytotoxic axotomy. In the present study, however, proliferation was not seen. The number of cells which were labeled by antibodies against HLA-DR, LCA, and CD 68 was within the range of that in normal brain (for rev. see [12]).

The results of our study have revealed a striking change of the microglial population later than 72 hours post TBI. After this time microglial cells increased in numbers binding HAM 56 and the lectin RCA-1. The pan-macrophage immunomarker HAM 56 is known to be also expressed in human alveolar macrophages. Its antigen is not exactly characterized yet. Hulette [13] has observed HAM 56-positivity only under pathological conditions. This would implicate that microglial cells have either changed to phagocytosing cells or that an increased number of blood-borne macrophages has invaded the brain. Although we have counted only those cells with a ramified or amoeboid morphology, this does not exclude invasion of blood-borne macrophages. The studies of Sievers [23] have demonstrated that macrophages have the capacity to

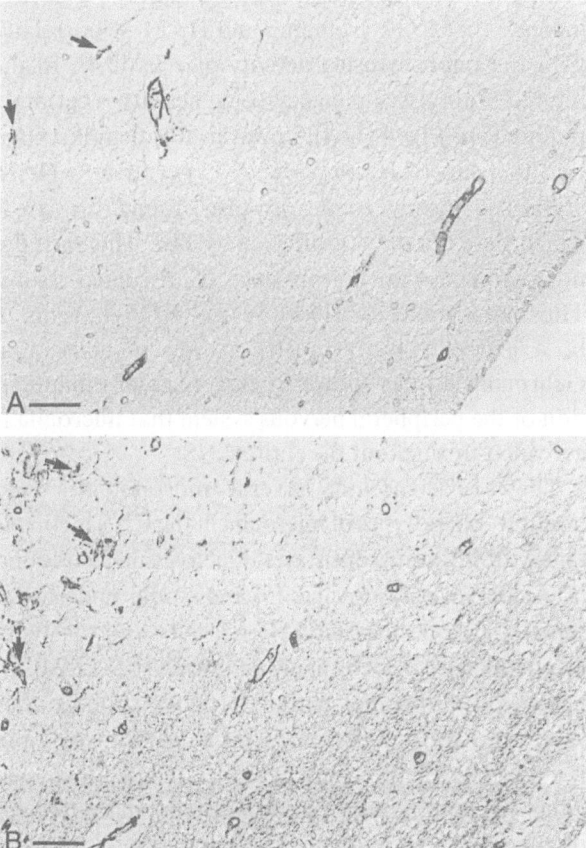

Fig. 6. Markers which were found on activated microglial cells in the later phase of TBI were only expressed in the deeper layers. Both micrographs demonstrate the area also shown in figure 5. HAM 56-positive (A) and RCA-1- positive cells (B) (arrows) were not present in the subependymal tissue. (bar = 40 μm)

acquire microglial properties in vitro based on both, morphology and electrophysiology. Astrocytes induce macrophages to express a typical inward rectifying ion-channel which is only found in microglia [15].

Occurrence of RCA-1 labeling not earlier than 72 hours post injury is a surprising result. Manoji [17] have described that the lectin can also label microglia in normal brain. He exclusively studied the brain of younger persons who apparently have not suffered from CNS diseases. It is believed that lectin binds specifically to β-galactose and β-N-acetylgalactos-amines. It might be possible that up to 72 hours post TBI, the absence of lectin binding is due to prolonged formalin fixation. This was, however, also true for those cases who survived TBI for more than 72 hours. The only possibility to explain this result is that after 72 hours specific galactose residues were increased on the cell surface. Due to the prolonged formalin fixation during the early phase the low number of galactose residues are not detectable by lectin binding before 72 hours after TBI.

Later galactosylation was most likely increased resulting in a positive lectin binding. Lines of evidence indicate that RCA-1 binding and HAM 56 correlates with the phagocytosing activity of cells [2,4]. In the injured brain tissue phagocytosing activity is certainly required. In Fig. 4 (D–I), however, it is demonstrated that the shift of HAM 56/RCA-1- negative to HAM 56/RCA-1-positive cells is also found in areas which were not directly affected by TBI. This shift did not correlate with the presence of disrupted axons. Thus, appearance of HAM 56 and RCA-1 seems to be a generally characteristic marker of activation. Gehrmann [8] has shown in experimental inflammation of the peripheral nervous system that microglia is activated throughout the entire CNS.

The coexistence of several morphologies and immuno-phenotypes of microglia in the CNS is consistent with the numerous states of activation also found in peripheral macrophages [5] and would underline a functional and antigenic heterogeneity and plasticity as postulated by Gehrmann and Streit [6,9,12,24]. In agreement with other authors [1,9] we suppose that there must be specific control mechanisms involved in the axonal reaction and neuronal cell death which either prevent microglia from becoming macrophages or conversely, promote their differentiation into macrophages. That macrophages can be changed into microglial cells is still considered [23]. There are cytokines which may contribute to the regulation of the functional role of microglial cells such as γ-inter-feron, interleukines, tumor necrosis factors, and some of the glial growth factors (for rev. see [20,21]). Yet, a definitive functional pattern and pathway of stimulating and inhibiting factors influencing activity of microglial cells in phagocytosis has not emerged so far.

Our results emphasize a delayed activation of microglial cells in TBI which is reflected by the expression of HAM 56 not earlier than 72 hours post injury. This delay does not correspond to microglial activation in experimental cerebral ischemia [1,6,9]. In these experiments microglial cells developed cytotoxic activity within 24 hours. Further investigation might elucidate whether or not the delayed activation of microglia can be used as a therapeutical window to reduce secondary brain damage from head injury.

Acknowledgement

We thank Gudrun Albrecht, Institute of Brain Research, University of Tübingen, Tübingen, FRG, for excellent preparation of the micrographs.

We thank also Dr. Norbert Benda, Institute for Medical Biometry, University of Tübingen, Tübingen, FRG, for supporting the statistical evaluations.

References

1. Banati RB, Gehrmann J, Schubert P, Kreutzberg GW (1993) Cytotoxicity of microglia. Glia 7: 111–118
2. Colton CA, Abel C, Patchett J, Keri J, Yao J (1992) Lectin staining of cultured CNS microglia. J Histochem Cytochem 40: 505–512
3. del Rio- Hortega P (1932) Microglia. In: Penfield W (ed) Cytology and cellular pathology of the nervous system, Vol 2. Hoeber, New York, pp 481–534
4. Esiri MM, Morris CS (1991) Immunocytochemical study of macrophages and microglial cells and extracellular matrix components in human CNS disease. J Neurol Sci 101: 59–72
5. Finch CE, Laping NJ, Morgan TE, Nichols NR, Pasinetti (1993) TGF-β1 is an organizer of reponse to neurodegeneration. J Cell Biochem 53: 314–322
6. Gehrmann J, Bonnekoh P, Miyazawa T, Hossmann KA, Kreutzberg GW (1992) Immunocytochemical study of an early microglial activation in ischemia J Cereb Blood Flow Metab 12: 257–269
7. Gehrmann J, Banati R, Kreutzberg GW (1993) Microglia in the immune surveillance of the brain: human microglia constitutively express HLA-DR molecules. J Neuroimmunol 48: 189–198
8. Gehrmann J, Gold R, Linington C, Lannes-Vieira J, Wekerle H, Kreutzberg GW (1993) Microglial involvement in experimental autoimmune inflammation of the central and peripheral nervous system. Glia 7: 50–59
9. Gehrmann J, Kreutzberg GW (1993) Monoclonal antibodies against macrophages/microglia: immunocytochemical studies of early microglial activation in experimental neuropathology. Clin Neuropathol 12: 301–305
10. Gentlemann S, Nash MJ, Sweeting CJ, Graham DI, Roberts GW (1993) β-Amyloid precursor protein (βAPP) as a marker for axonal injury after head injury. Neurosci Lett 160: 139–144

11. Grady MS, McLaughlin MR, Christman CW, Valadka AB, Fligner CL, Povlishock JT (1993) The use of antibodies targeted against the neurofilament subunits for the detection of diffuse axonal injury in humans. J Neuropathol Exp Neurol 52: 143–152

12. Graeber MB, Streit WJ (1990) Microglia: immune network in the brain. Brain Pathol 1: 2–5

13. Hulette CM, Downey BT, Burger PC (1992) Macrophage markers in diagnostic neuropathology. Am J Surg Pathol 16: 493–499

14. Itagaki S, McGeer PL, Akiyama H (1988) Presence of T-cytotoxic suppressor and leukocyte common antigen positive cells in Alzheimer's disease brain tissue. Neurosci Lett 91: 259–264

15. Kettenmann H, Hoppe D, Gottmann K, Banati R, Kreutzberg GW (1990) Cultured microglial cells have a distinct pattern of membrane channels different from peritoneal macrophages. J Neurosci Res 26: 278–287

16. Lawson LJ, Perry VH, Dri P, Gordon S (1990) Heterogeneity in the distribution and morphology of microglia in the normal, adult mouse brain. Neuroscience 39: 151–170

17. Mannoji H, Yeger H, Becker LE (1986) A specific histochemical marker (lectin ricinus communis agglutinin-1) for normal human microglia and application to routine histopathology. Acta Neuropathol 71: 341–343

18. McGeer PL, Itagaki S, McGeer EG (1988) Expression of the histocompatibility glycoprotein HLA-DR in neurological disease. Acta Neuropathol 76: 550–557

19. Nissl F (1894) Über die Untersuchungsmethode des Zentralorgans speziell zur Feststellung der Lokalisation der Nervenzellen. Zentralbl Nervenheilkunde Psychiatr 17: 337–334

20. Otero GC, Merrill JE (1994) Cytokine receptors on glial cells. Glia 11: 117–128

21. Perry VH, Andersson PB, Gordon S (1993) Macrophages and inflammation in the central nervous system. TINS 16: 268–273

22. Sasaki A, Nakazato Y (1992) The identity of cells expressing MHC class II antigens in normal and pathological human brain. Neuropathol Appl Neurobiol 18: 13–26

23. Sievers J, Parwaresch R, Wottke HU (1994) Blood monocytes and spleen macrophages differentiate into microglia-like cells on monolayers of astrocytes: morphology. Glia 12: 245–258

24. Streit WJ, Graeber MB (1993) Heterogeneity of microglial and perivascular cell populations: insights gained from the facial nucleus paradigm. Glia 7: 68–74

25. Thomas WE (1992) Brain macrophages: elevation of microglia and their functions. Brain Res 510: 154–157

26. Unterharnscheidt F (1993) Stadien der sogenannten Rindenprellungsherde. In: Doerr W, Seifert G (eds) Pathologie des Nervensystems VI. Springer, Berlin Heidelberg New York, Tokyo, pp 397–423

Correspondence: R. Meyermann, M.D., Institute of Brain Research, Calwerstr. 3, D-72076 Tübingen, Federal Republic of Germany.

Acta Neurochir (1996) [Suppl] 66: 96–102

Altered β-APP Metabolism After Head Injury and its Relationship to the Aetiology of Alzheimer's Disease

D. I. Graham[1], S. M. Gentleman[2], J. A. R. Nicoll[1], M. C. Royston[3], J. E. McKenzie[3], G. W. Roberts[4], and W. S. T. Griffin[5]

[1]Department of Neuropathology, University of Glasgow, [2]Department of Anatomy, Charing Cross and Westminster Medical School, London, [3]Department of Psychiatry, Charing Cross and Westminster Medical School, London, [4]Department of Molecular Neuropathology, SmithKline Beecham, Harlow, and [5]Departments of Paediatrics/Anatomy, Arkansas Children's Hospital Research Institute, Little Rock, U.S.A.

Summary

There is increasing evidence of a link between head injury and the subsequent onset of Alzheimer's disease. Deposits of amyloid β-protein (Aβ) are found not only in cases of dementia pugilistica but in some 30 % of patients dying after a single episode of severe head injury. Detailed clinicopathological studies have shown that Aβ deposition is most likely, but not exclusively, to occur, the older the patient at the time of injury, and if the injury is the result of a fall. Distribution studies have shown that the Aβ is widely deposited in the neocortex and there is no apparent association with any of the multiple primary or secondary pathologies of traumatic brain injury. There is an increased expression of β-APP particularly in the pre-α cells of the entorhinal cortex and in areas of axonal damage. Recent molecular genetic studies have shown that there is a strong association between deposits of Aβ and the apolipoprotein E genotype of the individual.

Keywords: Head injury; β-APP expression; deposition of Aβ.

Introduction

In 1993 we reviewed the epidemiological and neuropathological evidence that provided a link between a history of head injury and the subsequent development of Alzheimer's disease [11]. We hypothesised that, like other environmental precipitants of Alzheimer's disease, trauma probably acts through a variety of mechanisms that result in the overexpression of β-amyloid precursor protein (β-APP) and ultimately the deposition of Aβ. In the last few years this hypothesis has been tested by undertaking additional studies. However, before reviewing this recent work it would be appropriate to summarise some of the evidence that links a previous history of head injury with Alzheimer's disease.

Background

Head injury is an epidemiological risk factor for Alzheimer's disease [22,23,25]. From such studies it has been variously estimated that head injury plays a role in some 2–20 % of cases of the disease. Although suggestive these data are essentially circumstantial whereas considerable support for the hypothesis has been derived from a demonstration of Alzheimer-like pathology in the brains of boxers with dementia pugilistica apparently as a long term consequence of repeated blows to the head [33]. Dementia pugilistica is characterised neuropathologically by a cavum septum, loss of nerve cells, including depigmentation of the substantia nigra, scarring of the cerebellum, and widely distributed neurofibrillary tangle formation in the cortex [9]. Dementia pugilistica was always regarded as a separate diagnostic entity from Alzheimer's disease but the demonstration that large numbers of diffuse plaques composed of Aβ are present in the brains of boxers with dementia pugilistica raised the possibility that repeated blows to the head could trigger the deposition of Aβ in the brains of certain susceptible individuals [6,33,34]. Such evidence therefore strongly supported the view that Alzheimer-like pathology might develop in long term survivors after head injury. Pertinent questions therefore are how often and how quickly can these changes occur after head injury, what is the distribution of these changes, what may be the underlying metabolic abnormality, and is it possible to identify a susceptible population?

Amyloid β-Protein Deposition in the Brain After Severe Head Injury

In an earlier study of 16 patients aged 10–63 years who sustained head injury and survived for only 6–18 days, we showed using an antibody to Aβ extensive deposits of the protein in the cortex in 6 of the 16 patients – 38 % [35]. An even larger frequency (50 %) was found by Huber *et al.* [18]. Criticism of these findings suggested that the observations in these patients with severe head injury might be an artefact related either to the age of the patients or to an accumulation of Aβ containing macrophages surrounding the vascular lesions found in these patients. In order to test our hypothesis further that head injury can trigger Aβ deposition in the cortex we examined multiple cortical areas from 152 patients with an age range of 8 weeks to 81 years after a single episode of severe head injury with a survival time of between 4 hours and 2.5 years. The series was compared with a group of 44 neurologically normal controls aged 51–80 years. Immunostaining using the same specific antibody of Aβ, confirmed the original finding that about 30 % of cases of head injury have deposits of Aβ in one or more cortical areas and that such deposits could occur in patients as young as 10 years of age and with a survival time as short as 4 hours. Comparison of the rates of occurrence of multiple clinical features in the two groups showed that of the multiple variables studied there were only two statistically associated with the deposition of Aβ; first that the deposition of Aβ was more likely in patients over the age of 50 years with head injury compared with the age and sex matched controls (p < 0.01) and that the deposition of Aβ was more likely to be present in those patients who had been injured in a fall than those without (p < 0.02). The exact nature of these associations is not known but it is clearly of considerable interest in relation to the clinical outcome of patients after severe head injury [36].

What is the Distribution of Aβ Protein and Its Relationships to the Other Pathologies in the Brain Following Severe Head Injury?

In this study the distribution of Aβ was mapped in 14 patients aged 65 years or less and in whom it was known that the protein had been deposited after head injury and correlated with the various pathologies of traumatic brain injury. For this purpose immunohistochemistry was carried out on 44 representative areas of the cerebral and the cerebellar hemispheres and the brain stem. Contiguous blocks were taken from the left and right cerebral hemispheres at the level of the prefrontal lobes, the temporal lobes at their anterior, mid, and posterior extents and the occipital lobes. All sections were pretreated with 80 % formic acid for eight minutes, then incubated overnight with a monoclonal antibody to residues 8–17 of Aβ (Dako Limited, UK) at a dilution of 1 in 1,000. The sections were examined "blind" and all the histological abnormalities were recorded on a series of diagrams. This study showed that Aβ is widely distributed throughout the neocortex and that there is no correlation between its presence and that of cerebral contusions, intracranial haematoma, axonal injury, ischaemic brain damage, brain swelling or the pathology of raised intracranial pressure.

Although the amounts of Aβ deposited differed from case to case the pattern of deposition was constant in so far as it was widely distributed in grey matter but not in white matter. Furthermore, the Aβ deposits were usually bilateral and in the cortex there were equal amounts deposited in the cortex of the frontal, temporal, parietal and occipital lobes although Aβ was found more commonly in the cortex of the insula than elsewhere. Deposition appeared to be at random being present equally throughout the cortex of the cerebral hemispheres and without any accentuation or localisation that could be accounted for by various functionally related anatomical pathways. In contrast to the cortex, deposits were found much less commonly in the basal ganglia, hypothalamus and thalamus; rarely was it found in the globus pallidus [15].

It has been shown in the brains of patients with Downs Syndrome (trisomy 21) and in the early stages of Alzheimer's disease that the earliest pathological changes, including the deposition of Aβ, are to be found in the medial parts of the temporal lobes [3]. As already noted the deposition of Aβ is widely distributed and although present throughout the limbic system is also present in many other brain areas. If head injury is indeed a risk factor for the subsequent development of Alzheimer's disease then presumably, with increased survival, the processes by which Aβ is deposited are down regulated in brain areas other than within the limbic system.

What is the Source of the β-Amyloid β-Protein?

Increases in neuronal β-APP immunoreactivity are readily detectable in the temporal lobes of patients with Downs syndrome, Alzheimer's disease and epilepsy

[43]. No case in our series of head injury was found to have Aβ deposition in the absence of increased β-APP immunoreactivity. We therefore hypothesise that whereas Aβ deposition is contingent on increased amounts of neuronal β-APP it is by no means an inevitable consequence. On this basis we have hypothesised that an increase in β-APP, might enhance the production of Aβ, lead to the deposition of diffuse Aβ plaques and thus set the scene for the pathogenesis of Alzheimer's disease. In order to test this hypothesis we quantified using computerised imaging analysis the numbers of β-APP immunoreactive pre-alpha neurons (layers 2 of the entorhinal cortex) of 13 patients aged between 16 and 77 years who died within 28 days of a head injury, and compared them to 17 controls aged between 11 and 84 years who died as a result of a variety of conditions [24]. This study revealed a significant increase in the number of β-APP immunoreactive pre-alpha cells in the entorhinal cortex of patients who had died as a result of the head injury. These findings were consistent with our previous results [36] and with the observations of others reporting upregulation of β-APP expression in animal models of head injury [1,31,38,49]. This upregulation of β-APP might represent a normal protective response to neuronal stress or injury. However, failure subsequently to downregulate β-APP expression might lead to what is otherwise a normal restorative process becoming one that triggers a cascade that culminates in the neuropathology of Alzheimer's disease, by the overproduction or sustained production of β-APP which in turn may result in an increase in the amyloidogenic processing of β-APP a subsequent increase in Aβ and the deposition of Aβ in plaques. If such a cascade exists then an initial increase of β-APP production may be critical to the subsequent development of Alzheimer-type pathology.

What are the Possible Mechanisms Whereby the Upregulation of β-APP May Result in Alzheimer-Type Pathology?

There is increasing evidence that the expression of β-APP is a component of the brains' acute phase response to injury. In the central nervous system interleukin 1 (IL-1) is synthesised and secreted by microglia. Although IL-1 may contribute to repair functions as it activates astrocytes which release neurotrophic factors, such as the astrocytic cytokine S100β [21,26], excessive microglial expression of IL-1 has been implicated in the pathogenesis of Alzheimer's disease [17,37,45,52]. Given that IL-1 *in vitro* can induce the expression of β-APP [14] and the processing of β-APP via the secretory cleavage pathway [4] it was decided to study the involvement of activated microglia and IL-1 in the early stages following head injury. Seven patients aged between 23 and 65 years dying within 12 hours to 10 days following head injury and an equal number of suitably aged sex matched controls were studied. Using a double immunohistochemical labelling technique to localise IL-1α and β-APP in sections from the medial temporal lobe including the hippocampus, it was found that the number of activated microglia expressing this cytokine was increased along with their immunoreactive intensity. These changes were found throughout the medial part of the temporal lobe and were unrelated to other pathologies and this increased microglial IL-1α expression correlated both spatially and quantitatively with elevated β-APP expression by neurons and neurites. These findings are reminiscent of the changes found in Alzheimer's disease and suggested that the development of neuritic plaques and of Alzheimer's disease in long term survivors of head injury may be related to microglial IL-1α promotion of neurite growth and neuritic β-APP expression after head injury. The ability of IL-1 to induce excessive expression and stimulate processing of β-APP suggested that induced expression of IL-1 and β-APP may be linked, and in fact that the elevated levels of IL-1α may be directly responsible for the increased amounts of β-APP in neurons of head-injured patients. Moreover chronic overexpression of IL-1 may explain the long term increase in β-APP and its processing that may give rise to the later development of Alzheimer's disease.

These findings are consistent with the hypothesis that there is a threshold of prolonged acute phase activation beyond which IL-1 is chronically elevated thus promoting a self sustaining and pathological positive feedback loop. If early events such as head injury precipitate Alzheimer's disease a potent acute phase molecule is necessarily involved in such a circle and we propose that IL-1 is such a molecule and that chronically elevated levels of the protein ultimately lead to neurodegenerative changes [17].

Is There a Genetic Susceptibility to the Effects of a Head Injury?

The fact that the deposition of Aβ is found in only 30 % of patients following fatal head injury suggests that there may be a genetic susceptibility to the effects of head injury. Some genetic abnormalities are causative for Alzheimer's disease, the most common of which is Downs Syndrome with trisomy 21 [48]. Much less common are mutations in the β-APP gene encoded on chromosome 21 [12,13,19]. It is a mismetabolism of this protein which gives rise to the Aβ found in the diffuse plaques. Recently abnormalities on chromosome 14 [40] and chromosome 19 [30] have been linked to certain cases of familial Alzheimer's disease [51].

The gene for apolipoprotein E (ApoE) located on the long arm of chromosome 19, has been linked with Alzheimer's disease [30,41]. Further confirmation was obtained by Corder *et al.* [7] who showed that the ε4 allele of ApoE was approximately 50 % in patients with late onset of Alzheimer's disease compared with 15 % of controls. The association between ApoE and Alzheimer's disease has now been confirmed for both sporadic and familial cases [5,8,39,41,49] and there are now reports that early onset disease is also associated with an even higher frequency of the ε4 allele [10,29,53].

Given this, and the observation that head injury can trigger Alzheimer-like pathology, we decided to investigate whether this environmental event would show an association or interaction with known genetic risk factors for Alzheimer's disease. Such an association could explain why Aβ deposition occurs in only some 30 % following head injury. ApoE genotype was determined on formalin fixed paraffin embedded tissue prepared for the polymerase chain reaction on 90 patients who died within 2 weeks of a head injury. Of the 23 cases in whom Aβ diffuse plaques were identified, the frequency of ApoE ε4 was 0.52 compared with 0.16 for those patients without Aβ deposits (P < 0.0001). Stratification of the results by patient age confirmed that the relationship between ApoEε4 and Aβ deposition held for those under 60 years of age, below which age Aβ deposits were not seen in controls who had not had a head injury. Such a relationship also held when the head-injured patients under the age of 50 years were considered as a separate group, the ApoEε4 frequency being 0.39 compared with 0.18 for the Aβ absent cases. Furthermore, there was a gene dose effect, the proportion of head-injured patients with Aβ deposits increasing with the number of Apoε4 alleles from 10 % for those without an ε4 allele to 35 % for those with one ε4 allele to 100 % for Eε4 homozygotes [28]. These results support the contention that ApoE isoform E4 or alternatively absence of isoforms E2 and E3 influences βAPP/Aβ metabolism in such a way as to augment the formation of Aβ. Our findings indicate that this overexpression is considerably more likely to

result in deposition of Aβ in those individuals possessing an ApoE ε4 allele: 45 % of those with one or more ε4 alleles are Aβ positive compared with 10 % of those with no ε4 allele. Recent epidemiological data from the study of risk factors in patients with Alzheimer's disease suggests that there is a powerful synergistic interaction between possession of ApoEε4 and the effects of a head injury [23].

Discussion

Since our original study [35] we have not only replicated but also extended the original finding that deposits of Aβ are present in the brains of about 30 % of patients dying after a single episode of severe head injury and that diffuse deposits of Aβ may occur in young patients within a matter of a few hours of the injury. However, there is an effect of age in that the older the patient at the time of head injury the greater is the likelihood that Aβ will be deposited in the brain. We have argued that a preponderance of diffuse plaques such as that seen in the cortex in dementia pugilistica might indicate a long term consequence of head injury [6] and by extrapolation it has also been suggested that a large ratio of diffuse to classic plaques might indicate an environmental trigger to the Alzheimer's disease process. It has been assumed previously that the process of Aβ deposition and subsequent plaque formation that characterises Alzheimer's disease is a gradual one, but our observations would suggest that this need not necessarily be so, and that Aβ deposition can occur within hours of head injury.

When taken in conjunction with the data from animal studies [20,27,31,32,42], our studies in fatal human head injury suggest that the induction of β-APP 751/770 in the brain is a normal response to neuronal stress and that in some 30 % of the head-injured patients there is deposition of Aβ in individuals whose susceptibility is closely associated with apolipoprotein E genotype. Both the genetic and immunohistological studies in head injury reinforce the point that whatever the cause of Alzheimer's disease the overexpression and abnormal metabolism β-APP and consequent deposition of Aβ is an essential event in the pathogenesis and evolution of the disease. The mechanism by which this acute phase response is perpetuated into a cascade that ultimately terminates in Alzheimer's disease is not clear, although, evidence is accumulating in Downs Syndrome, Alzheimer's disease, head injury and epilepsy to suggest that the process is related to the activation of microglia and the induction of IL-1 cytokine.

Evidence of irreversible ischaemic brain damage is present in over 90 % of fatal head injuries [16]. Although the mechanisms of this ischaemic brain damage are complex it is likely that the excitatory amino acids glutamate and aspartate play a central role in receptor mediated neuronal damage. *In vivo* microdialysis studies of brain injury in both man and experimental animals have shown a large transient increase in various excitatory amino acids which in turn trigger a cascade of molecular events including the increased expression of β-APP and possible irreversible ischaemic brain damage. It was therefore surprising that an association between hypoxia/ischaemia and the deposition of Aβ was not found in a large group of patients dying after head injury [36].

That changes in cerebral perfusion pressure might be a risk factor for the development of plaques in non-demented subjects with critical coronary artery disease compared to non heart disease controls was reported by Sparks *et al.* [44]. The increased numbers of plaques were found in the arterial boundary zones and it was suggested that their localisation was a product of fluctuations in the cerebral perfusion pressure as a result of cerebral oligaemia induced by coronary artery disease. Epidemiological evidence has also suggested that there is an increased risk of dementia after stroke [2,46]. In this study it was concluded that ischaemic stroke in elderly patients increases the long term risk of developing dementia by approximately five fold.

Several population studies have shown that ApoE polymorphism contributes to variations in plasma lipid and lipoprotein concentrations and that $\varepsilon4$ allele bearing subjects consistently have higher total low density lipoprotein cholesterol levels [50]. The $\varepsilon4$ allele predisposes to coronary artery disease and probably by analogy also to cerebrovascular disease. In a study of 187 patients with a probable or possible clinical diagnosis of various forms of dementia and in 156 autopsied patients with dementia it was concluded that ApoE is a major risk factor for Alzheimer's disease alone or in conjunction with other pathologies. The situation with patients who had had a stroke therefore seems analogous to patients dying from head injury in that when either of these two diseases occurs in an individual with an $\varepsilon4$ allele there is an increased risk of subsequent development of Alzheimer's disease.

Conclusions

The deposition of Amyloid β-protein in the brain plays a key role in the pathogenesis of Alzheimer's disease. Head injury is an epidemiological risk factor for Alzheimer's disease and deposition of Aβ occurs in the brain of about 30 % of fatal head injuries. The frequency of ApoE$\varepsilon4$ in the Aβ positive cases of fatal head injury is higher than in most studies of Alzheimer's disease. This finding provides further evidence linking ApoE$\varepsilon4$ with Aβ deposition *in vivo* and suggests that known environmental and genetic risk factors for Alzheimer's disease may act additively.

References

1. Abe K, Tanzi RE, Kogure K (1991) Selective induction of Kunitz-type protein: inhibitor domain containing amyloid precursor protein in RNA after persistent focal ischaemia in rat cerebral cortex. Neurosci Lett 125: 172–174
2. Betard C, Robitaille Y, Gee M, Tiberghien D, Larrivee D, Roy P, Mortimer JA, Gauvreau D (1994) Apo E allele frequencies in Alzheimer's disease, Lewy body dementia, Alzheimer's disease with cerebrovascular disease and vascular dementia. Neuroreport 5: 1893–1896
3. Braak H, Braak E (1991) Neuropathological staging of Alzheimer-related changes. Acta Neuropathol 82: 239–259
4. Buxbaum JD, Oishi M, Chem HI, Pinkas-Kramarski R, Jaffe EA, Gandy SE, Greengard P (1992) Cholinergic agonists and interleukin 1 regulate processing and secretion of the Alzheimer β/A4 amyloid protein precursor. Proc Nat Acad Sci USA 89: 10075–10078
5. Chartier-Harlin M-C, Parfitt M, Legrain S, Perez-Tur J, Brousseau T, Evans A, Berr C, Vidal O, Roques P, Gourlet V, Fruchart JC, Delacourte A, Rosser M, A Mouyel P (1994) Apolipoprotein E4 alelle as a major risk factor for sporadic early- and late-onset forms of Alzheimer's disease: analysis of the 19q 13.2 chromosomal region. Hum Mol Genet 3: 569–574
6. Clinton J, Ambler MW, Roberts GW (1991) Post-traumatic Alzheimer's disease: preponderance of a single plaque type. Neuropathol Appl Neurobiol 17: 69–74
7. Corder EH, Saunders AM, Strittmatter WJ, Schmechel DE, Gaskell PC, Small GW, Roses AD, Haines JL, Pericak Vance MA (1993) Gene dose of apolipoprotein E type 4 allele and the risk of Alzheimer's disease in late onset families. Science 261: 921–923
8. Corder EH, Saunders AM, Risch NJ, Strittmatter WJ, Schmechel DE, Gaskell PC, Rimmeler JB, Locke PA, Conneally PM, Schmader KE, Small GW, Roses AD, Haines JL, Pericak Vance MA (1994) Protective effect of apolipoprotein E type 2 allele for late onset Alzheimer disease. Nature Genet 7: 180–184
9. Corsellis JAN, Bruton CJ, Freeman-Browne D (1973) The aftermath of boxing. Psychol Med 3: 270–273
10. Dai XY, Nanko S, Hattori M, Fukuda R, Nagata K, Isse K, Uski A, Kazamatsuri H (1994) Association of apolipoprotein E4 with sporadic Alzheimer's disease is more pronounced in early onset type. Neurosci Lett 175: 74–76
11. Gentleman SM, Graham DI, Roberts GW (1993) Molecular pathology of head injury: altered β-APP metabolism and the aetiology of Alzheimer's disease. Prog Brain Res 96: 237–246
12. Goate A, Chartier-Harlin M-C, Mullan M (1991) Segregation of a missense mutation in the amyloid precursor protein gene with familial Alzheimer's disease. Nature 349: 704–706
13. Goldgaber D, Lerman MI, McBride OW, Saffiotti U, Gajdusek DC (1987) Characterization and chromosomal localization of a DNA including brain amyloid of Alzheimer's disease. Science 235: 877–880
14. Goldgaber D, Harris HW, Hla T, Maciag T, Donnelly RG, Jacobsen JS, Vitek MP, Gajdusek DC (1989) Interleukin 1 regu-

lates synthesis of amyloid β protein precursor mRNA in human endothelial cells. Proc Nat Acad Sci USA 86: 7606–7610

15. Graham DI, Gentleman SM, Lynch A, Roberts GW (1995) Distribution of β- amyloid protein in the brain following severe head injury. Neuropathol Appl Neurobiol 21: 27–34

16. Graham DI, Ford I, Adams JH, Doyle D, Teasdale GM, Lawrence AE, McLellan DR (1989) Ischaemic brain damage is still common in fatal non-missile head injury. J Neurol Neurosurg Psychiatry 52: 346–350

17. Griffin WST, Shen JG, Gentleman SM, Graham DI, Mrak RE, Roberts GW (1994) Microglial Interleukin-1α expression in human head injury: correlations with neuronal and neuritic β-amyloid precursor protein expression. Neurosci Lett 176: 133–136

18. Huber A, Gabbert K, Keleman J, Cervos-Navarro J (1993) Density of amyloid plaques in brain after head trauma. J Neurotrauma 10 [Suppl 1]: S180

19. Kang J, Lemaire HG, Unterbeck A, Salbaum JM, Masters CL, Grzeschik KH, Multhaup G, Beyreuther K, Muller-Hill B (1987) The precursor of Alzheimer's disease amyloid A4 protein resembles a cell-surface receptor. Nature 325: 733–736

20. Kawarbayashi T, Shoji M, Harigaya Y, Yamaguchi H, Hirai S (1991) Expression of APP in the early stages of brain damage. Brain Res 563: 334–338

21. Marshak DR, Pesce SA, Stanley LC, Griffin WST (1991) Increased S100β-neurotrophic activity in Alzheimer's disease temporal lobe. Neurobiol Aging 13: 1–7

22. Mayeux R, Ottman R, Tang M-X, Noboabauza L, Marder K, Gurland B, Stern Y (1993), Genetic susceptibility and head injury as risk factors for Alzheimer's disease amongst community-dwelling elderly persons and their first-degree relatives. Ann Neurol 33: 494–501

23. Mayeux R, Ottman R, Maestre G, Ngae BS, Tang M-X, Ginsburg H, Chun M, Tycko B, Shelanski M, Synergistic effects of traumatic head injury and apolipoprotein-ε in patients with Alzheimer's disease. Neurology 45: 555–557

24. McKenzie JE, Gentleman SM, Roberts GW, Graham DI, Royston MC Increased numbers of β-APP immunoactive neurons in the entorhinal cortex after head injury. Neuroreport 6: 161–164

25. Mortimer JA, Van Duijn CM, Chandra V (1991) Head trauma as a risk factor for Alzheimer's disease: a collaborative re-analysis of case-control studies. Int. J Epidemiol 20 [Suppl 2]: S28–35

26. Mrak RE, Sheng JE, Griffin WST (1995) Glial cytokines in Alzheimer's disease. Review and pathogenetic implications. Hum Pathol 26: 816–823

27. Nakamura Y, Takeda M, Niigawa H, Hariguchi S, Nishimura T (1992) Amyloid β-protein precursor deposition in rat hippocampus lesioned by ibotenic acid injection. Neurosci Lett 136: 95–98

28. Nicoll JAR, Roberts GW, Graham DI (1995) Apolipoprotein Eε4 allele is associated with deposition of amyloid β-protein following head injury. Nature Medicine 1: 135–137

29. Okuizumi K, Onodera O, Tanaka H, Kobayashi H, Tsuji S, Takahashi H, Oyanagi K, Seki K, Tanaka M, Naruse S, Miyatake T, Mizusawa H, Kanazawa I (1994) Apoe-4 and early onset Alzheimer's. Nature Genet 7: 10–11

30. Pericak-Vance MA, Bebout JL, Gaskell PC, Yamaoka LH, Hung WY, Alberts MJ, Walker AP, Bartlett RJ, Haynes CA, Welsh KA, Earl NL, Heyman A, Clark CM, Roses AD (1991) Linkage studies in familial Alzheimer's disease: evidence for chromosome 19 linkage. Am J Hum Genet 48: 1034–1050

31. Pluta R, Kida E, Lossinsky AS, Golabak AA, Mossakowski MJ, Wisniewski HM (1994) Complete cerebral ischaemia with short-term survival in rats induced by cardiac arrest. Extracellular accumulation of Alzheimer's β-amyloid protein precursor in the brain. Brain Res 649: 323–328

32. Quon C, Wang Y, Catalano R, Marian Scardina J, Murakami K, Cordell B (1991) Formation of β-amyloid protein deposits in brains of transgenic mice. Nature 352: 239–241

33. Roberts GW, Allsop D, Bruton CJ (1990) The occult aftermath of boxing. J Neurol Neurosurg Psychiatry 53: 373–378

34. Roberts GW, Whitwell HL, Acland PR, Bruton CJ (1990) Dementia in a punch-drunk wife. Lancet 335: 918–919

35. Roberts GW, Gentleman SM, Lynch A, Graham DI (1991) βA4 amyloid protein deposition in brain after head trauma. Lancet 338: 1422–1423

36. Roberts GW, Gentleman SM, Lynch A, Murray L, Landon M, Graham DI (1994) βA4 amyloid protein deposition in brain after severe head injury: implications for the pathogenesis of Alzheimer's disease. J Neurol Neurosurg Psychiatry 57: 419–425

37. Rogers J, Cooper NR, Webster S, Schultz J, McGeer PL, Styren SD, Civin WH, Brachova L, Bradt B, Ward P (1992) Complement activation by β-Amyloid in Alzheimer's disease. Proc Nat Acad Sci 89: 10016–10020

38. Royston MC, Rothwell NJ, Roberts GW (1992) Alzheimer's disease: aetiology: pathology to potential treatments. Trends Pharmacol Sci 13: 131–133

39. Saunders AM, Strittmatter WJ, Schmechel D, St George-Hyslop PH, Pericak-Vance MA, Joo SH, Rosi BL, Gusella JF, Crapper Maclachlan DR, Alberts MJ, Hulette C, Crain B, Goldgaber D, Roses AD (1993) Association of apolipoprotein E allele 4 with late-onset familial and sporadic Alzheimer's disease. Neurology 43: 1467–1472

40. Schellenberg GD, Bird TD, Wijsman EM, Orr HT, Anderson L, Nemens E, White JA, Bonny-Castle L, Weber JL, Alonso ME, Potter H, Heston LL, Martin GM (1992) Genetic linkage evidence for a familial Alzheimer's disease/locus on chromosome 14. Science 258: 668–671

41. Yu C-E, Payami H, Olson JM, Boehnke M, Wijsman EM, Orr HT, Kukull WA, Goddard KAB, Nemens E, White JA, Alonso ME, Taylor TD, Ball MJ, Kaye J, Morris J, Chui H, Sadovnick AD, Martin GM, Larson EB, Heston LL, Bird TD, Schellenberg GD (1994) The Apolipoprotein E/C1/C11 gene cluster and late-onset Alzheimer's disease. Am J Hum Genet 54: 631–642

42. Scott JN, Parhad IM, Clark AW (1991) β-Amyloid precursor protein gene is differentially expressed in axonotomized sensory and motor systems. Mol Brain Res 10: 315–325

43. Sheng JE, Boup FA, Mrak RE, Griffin WST (1994) Increased neuronal Beta-amyloid precursor protein expression in human temporal lobe epilepsy: association with interleukin – 1 alpha immunoreactivity. J Neurochem 63: 1872–1879

44. Sparks DL, Liu H, Scheff SW, Coyne CM, Hunsaker JC (1993) Temporal sequence of plaque formation in the cerebral cortex of non-dementia individuals. J Neuropathol Exp Neurol 52: 135–142

45. Stanley LC, Griffin WST (1990) Localization of 1L-1α and 1L-1β in diseases with gliosis, dementia and immune suppression. Soc Neurosci 16: 1345

46. Tatemichi TK, Paik M, Bagiella E, Desmond DW, Stern Y, Sano M, Hauser WA, Mayeux R (1994) Risk of dementia after stroke in a hospitalized cohort: Results of a longitudinal study. Neurology 44: 1885–1891

47. Tomimoto H, Wakita H, Akiguchi I, Nakamura S, Kimura J (1994) Temporal profiles of accumulation of amyloid βA4 protein precursor in the gerbil after ischaemic stress. J Cereb Blood Flow Metab 14: 565–573

48. Tomlinson BE (1992) Aging and the Dementias. In: Adams JH, Duncan LW (Eds), Greenfields neuropathology, 5th Ed. Edward Arnold, London, pp 1284–1410

49. Tsai M-S, Tangalos EG, Peterson RC, Smith GE, Schaid DJ, Kokmen E, Ivnik RJ, Thibodeau SN (1994) Apolipoprotein E: risk factor for Alzheimer's disease. Am J Hum Genet 6: 643–649

50. Utermann G, Kindermann I, Kaffarnik H, Steinmetz A (1984) Apolipoprotein E phenotypes and hyperlipidemia. Hum Genet 65: 232–236

51. Van Broeckhoven CL (1995) Molecular genetics of Alzheimer disease: identification of genes and gene mutations. Europ Neurol 35: 8–19

52. Van Denabelle P, Fiers W (1991) Is amyloidogenesis during Alzheimer's disease due to an 1L-1/1L-6- mediates "acute phase response" in the brain? Immunol Today 12: 217–219

53. Van Duijn CM, de Kinijff P, Cruts M, Wehnert A, Havekes LM, Hofman A, Van Broeckhoven C (1994) Apolipoprotein E4 allele in a population-based study of early-onset Alzheimer's disease. Nature Genet 7: 74–78

Correspondence: D. I. Graham, M.D., Department of Neuropathology, Institute of Neurological Sciences, Southern General Hospital, Glasgow, G51 4TF, U.K.

Acta Neurochir (1996) [Suppl] 66: 103–106

Principles of Neuronal Regeneration

G. W. Kreutzberg

Department of Neuromorphology, Max-Planck-Institute of Psychiatry, Martinsried, Federal Republic of Germany

Summary

Studies of retrograde changes in axotomized motoneurons have revealed fundamental changes in morphology, metabolism and physiology of these cells. Restructuring of the granular endoplasmic reticulum, s.c. chromatolysis, seems to be the basis for increased and modified protein synthesis. While cytoskeletal proteins with the exception of the neurofilament triplet go up, enzymes and receptors related to neurotransmission go down and new growth associated proteins appear. There is an enhanced glucose uptake and iron metabolism. Complex changes in axonal transport have been observed. This may reflect the efforts of the regenerating nerve cell to compensate for its lost axon.

Keywords: Facial nucleus; glia; immediate early genes; growth factors.

Immediate Early Genes

Facial motoneurons respond with a number of molecular changes to peripheral transection of the facial nerve. In order to obtain insight into the transcriptional regulation during neuronal regeneration the expression of a number of immediate early genes (IEGs) was investigated after facial nerve lesion [4]. Some IEGs (such as c-fos, c-jun or jun B) are known to regulate gene expression. Quantitative Northern blot analysis covering a post-operative time course of hours, days and weeks revealed that axotomy produces a unique pattern of IEG induction in the facial nucleus, characterized by a long-term increase of the respective mRNAs. c-Jun and jun B RNAs, also present in low amounts in the unoperated nucleus, were strongly induced after injury. The increase of jun B and c-jun mRNA levels was detectable at 5 hours and these levels were maintained up to eight days after axotomy. c-Fos, however, known to act in concert with c-jun in other systems, was not expressed at basal levels in the unoperated nucleus, nor was c-fos mRNA induced by axotomy at all time points studied. Two members of the TIS family of immediate early genes, TIS 7 and TIS 11 mRNA, however, were detectable at low levels in normal facial nucleus. TIS 7 mRNA levels were unaffected by lesions, whereas TIS 11 mRNA levels were increased in a similar fashion as c-jun and jun B with an early rise at 5 h that lasted until day eight. With regard to the cellular distribution of these IEGs *in situ* hybridization histochemistry revealed that these mRNAs were present at basal levels in motoneurons and were exclusively induced in these cells in response to axotomy. Jun B mRNA, however, seemed to be present in motoneurons and glial cells under normal condition and after lesion; jun B therefore could be involved in the transcriptional regulation of both cell types.

Calcitonin Gene-Related Peptide (CGRP), a Putative Signal in Regeneration

Neuronal injury leads not only to changes in the neurons themselves but also in non-neuronal cells in their vicinity. Lesion of the facial nerve, for instance, results in a response of both astrocytes and microglia around the neuronal cell bodies in the facial nucleus. These glial responses have been known for some time, little information has been available on the signals that mediate neuron-glial communication after neuronal injury and during regeneration. Various classes of substances might be considered to be candidates as messengers mediating intercellular communication during regeneration including growth factors, cytokines and neuropeptides. Recent evidence suggests that calcitonin gene-related peptide (CGRP), a 37 amino acid neuropeptide, might play a role in this context [2].

After facial nerve transection, increases in CGRP immunoreactivity and CGRP mRNA are observed in facial motoneurons as early as 15 h after axotomy. Approximately 50 % of the facial nucleus motoneurons responded with an increase in CGRP mRNA as judged by in situ hybridisation histochemistry. Further analysis revealed a biphasic pattern of the CGRP increase with a first peak around day 3 and a second peak at day 21 after facial nerve transection. An increase in the biosynthesis of CGRP therefore occurs in response to motoneuron injury. In contrast, sciatic nerve transection led to a decrease in CGRP levels and CGRP mRNA in the dorsal root ganglion. Thus CGRP synthesis is differentially regulated in motor and sensory neurons in response to injury.

The early and dramatic increase of CGRP in injured facial motoneurons is one of the first observations concerning an increase in a putative intercellular messenger during nerve regeneration. Thus this peptide might play a role in the cellular reactions accompanying neuronal regeneration. *In vitro* studies demonstrated that CGRP has several effects on cultured astrocytes. Addition of CGRP to cultures of astrocytes from neonatal rat brain led to a change in morphology, flat polyglonal cells taking on a multipolar form with many processes. In addition, a stimulation of cyclic AMP accumulation was observed in these cells in response to CGRP indicating the presence of receptors

Table 1. *Mechanisms elicited in Regenerating Neurons of the Spinal Cord or Medulla in Response to Axonal Injury*

1. Synaptic changes
   - Shedding of presynaptic boutons from soma and stem dendrites (synaptic stripping by microglia)
   - Insulation of motoneurons by astroglial lamellae (chronic)
   - Retraction of dendrites leading to a reduction of the dendritic field by ca 24 – 34 %
   - Budding and sprouting of pathological dendraxons from neuronal soma and dendrites

2. Biochemical changes
   - Increase in glucose consumption
   - Increase in NO synthase
   - Increase in tubulin and actin
   - Decrease in neurofilament proteins
   - Enzymes of the neurotransmitter synthesis or degradation are down regulated (CAT, AChE)
   - Receptors for neurotransmitters are down regulated
   - Neuropeptides show changes (CGRP increase)

3. Changes in neurotrophic factors, growth associated proteins and their receptors
   - CNTF
   - NGF
   - Transferrin receptor
   - PDGF

for the peptide on astrocytes. An action of CGRP also at the transcriptional level in astrocytes was indicated by a rapid and strong induction of the c-fos proto-oncogene. These data indicate the ability of cultured astrocytes to respond to CGRP and thus open the possibility of a role for this peptide as mediator of neuron-glial interaction during neuronal regeneration.

CNTF, a Survival Factor for Lesioned Neurons

The response of nerve cell bodies to axonal injury is aimed at regeneration. However, in some systems, e.g. the vagal motor nucleus, or at some developmental stages, e.g. in the newborn, axotomized motoneurons degenerate. In a previous *in vivo* study it has been demonstrated that cell death of facial motoneurons can be prevented in newborn rats by CNTF, the ciliary neurotrophic factor. The factor was applied locally to the lesioned nerve in a piece of soaked foam. On the untreated side 80 % of the axotomized neurons showed severe retrograde changes and died within one week. Treatment with CNTF prevented neuronal death and suppressed chromatolysis, i.e. the motoneurons showed normal morphology [10]. The postnatal vulnerability of motoneurons is probably due to a lack of physiological CNTF production in the lesioned nerve. In contrast, Schwann cells of the adult peripheral nerve synthesize sufficient CNTF to keep motoneurons alive when axotomized. For the motoneuron CNTF is the first survival and lesion factor demonstrated *in vivo*.

Growth Factor Receptors

Interesting molecular changes with regard to growth factor receptors occur at the site of lesion in the peripheral nerve [9]. There is an early induction of specific PDGF-receptor on fibroblast-like cells. Initially a symmetrical appearance of PDGF-R is seen in the proximal and distal stump. Within 2 d p.op. receptor binding spreads throughout the distal stump. NGF-receptors show an early accumulation in the cut axons suggesting an axonal transport. The accumulation of NGF-R sites at later states is related to Schwann cells which are likely to become the extraneuronal source for NGF-R.

NGF is a well known neurotrophic factor for sensory and sympathetic neurons. It is produced in their target tissues, bound and taken up by NGF receptor-bearing sensory and sympathetic axon terminals and then transported retrogradely as a NGF-NGFR complex to their perikarya where it exerts its trophic function. We have studied changes in this axonal transport of endogeneous NGF, NGFR and NGFR saturation (NGF/NGFR) following axotomy and during regeneration of the rat sciatic nerve using NGF-ELISA and quantitative in situ NGFR autoradiography.

The retrograde transport of NGF decreased dramatically to 10 % of the normal levels one day after axotomy but then returned to a stable plateau of 30–37 % on day 3–13. The retrograde transport of NGFR decreased more gradually, reaching a similar plateau of approximately 40 % 3–13 days post op. Starting with day 21 the axonal transport of both endogeneous NGF and its receptor gradually increased, reaching normal levels at day 45. Interestingly, the NGFR saturation dropped precipitously to 15 % of the normal values 1 day p. op. but then rapidly recovered to normal levels (70–130 %) during the whole course of nerve regeneration. This normal NGFR saturation suggests that the endoneurium of the regenerating sciatic nerve provides the regenerating axons with NGF levels very similar to those in the normal peripheral target tissues. Despite this apparent ability of the axotomized endoneurium to replace the periphery, there is a 3-fold reduction in the axonal transport of NGF due primarily to the strong decrease in the neuronal sensitivity to the neurotrophic effects of NGF during regeneration.

Glial Reactions Accompanying Neuronal Response

Studies of the cellular reactions occurring in the facial nucleus following peripheral nerve transection have revealed microglia and astrocyte activation in a graded fashion [6]. In this remote lesion model the blood brain barrier is untouched. Thus, hematogenous cells are not invading this brain stem nucleus. Microglia, but not astrocytes undergo mitosis. They are or become positive for several markers, e.g. enzymes, lectins and immunomolecules. Within days motoneurons are covered by microglia and the presynaptic terminals are stripped from the soma and the stem dendrites. Such reactive microglia cells develop a cytoskeleton rich in vimentin. They can be stained by the lectin GSA I-B4 and show strong 5'-nucleotidase activity. Although the retrograde reaction is produced by a sterile lesion microglia express a number of immunomolecules, e.g. the complement receptor and MHC antigens class I and occasionally class II (Ia). A number of monocyte/macrophage antigens were absent from resting and activated microglia, e.g. ED-2.

Transformation of microglial cells from a quiescent or resting state to an activated state occurs within hours after injury. MUC 102, an antibody recognizing microglia and brain macrophages is a very sensitive marker of this activation [3], as is the amyloid precursor protein L-APP and the complement receptor CR3bi (Ox 42).

By experimental manipulation leading to retrograde nerve cell destruction microglia develop into brain macrophages with neuronophagic activity [1]. In the toxic ricin paradigm the nature of these microglia derived brain macrophages has been shown to be different from common blood-borne macrophages.

The repertoire of regulation of significant immunomolecules in activated microglia is not specifically related to regeneration only. In contrast, regulation seems to be essentially the same in all cases of microglial activation, whether occurring in trauma, ischemia, inflammation, autoimmune diseases, intoxication or even cortical spreading depression. The microglia seems to act as a multi-purpose defence cell of the parenchyma. It is activated quickly and augmented by mitosis, is an antigen presenting cell, has cytotoxic potential, is mobile and responsive to intercellular communication molecules and to changes in the ionic milieu and finally can become a brain macrophage [1,6,7]. During the axotomy response in the facial nucleus astrocytes become reactive, show

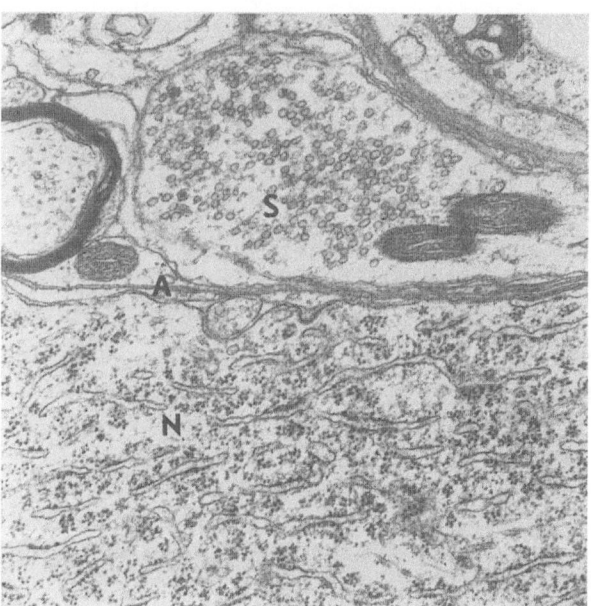

Fig. 1. 30 d after axotomy a facial motoneuron (N) exhibits a hypertrophic Nissl body with wide cisternae and a great number of ribosomes and polysomes. The neuron is covered by astroglial (A) processes. The large synaptic terminal (S) is separated from the neuron (synaptic stripping). Magnification 1 : 63000

enhanced GFAP synthesis and increased glial filaments and processes. They form lamellar stacks around the regenerating motoneurons insulating them from synaptic afferences and the proliferating perineuronal microglia (Fig. 1). Wrapping of neurons by astroglial lamellar stacks has been observed even after successful peripheral reinnervation, e.g. one year p.op [6]. Astrocytes could be the source of colony stimulating factors (CSF) acting as mitogens for microglia which increasingly express CSF receptors [8]. Cytokines such as IL-6 and TGF-beta have also been demonstrated to increase in reactive glial cells during the axotomy response in the facial nucleus [5].

Some neurons, e.g. in the dorsal vagal nucleus of the guinea pig grow new pseudo-dendritic processes (dendraxons) out of the soma which tend to surround the neuronal cell bodies. From these observations it is concluded that the regeneration process imposes profound changes on the neurons with regard to their cellular and synaptic organisation. These changes may also explain deficits in fine motor control in patients suffering, e.g. from a facial nerve trauma.

References

1. Banati RB, Gehrmann J, Schubert P, Kreutzberg GW (1993) Cytoxicity of microglia. Glia 7: 111–118

2. Dumoulin FL, Raivich G, Haas CA, Lazar P, Reddington M, Streit WJ, Kreutzberg GW (1992) Calcitonin gene-related peptide and peripheral nerve regeneration. Ann NY Acad Sci 657: 351–360
3. Gehrmann J, Kreutzberg GW (1991) Characterisation of two new monoclonal antibodies directed against rat microglia. J Comp Neurol 313: 409–430
4. Haas CA, Donath C, Kreutzberg GW (1993) Differential expression of immediate early genes after transection of the facial nerve. Neuroscience 53: 91–99
5. Kiefer R, Lindholm D, Kreutzberg GW (1993) Interleukin-6 and transforming growth factor-β1 mRNAs are induced in rat facial nucleus following motoneuron axotomy. Eur J Neurosci 5: 775–781
6. Kreutzberg GW (1993) Perineuronal glial reactions in regeneration of motoneurons. In: Fedoroff S, Juurlink BHJ, Doucette R (eds) Biology and pathology of astrocyte-neuron interactions. Altschul Symposia, vol 2. Plenum, New York, pp 283–290
7. Moneta ME, Gehrmann J, Töpper R, Banati RB Kreutzberg GW (1993) Cell adhesion molecule expression in the regenerating rat facial nucleus. J Neuroimmunol 45: 203–206
8. Raivich G, Gehrmann J, Kreutzberg GW (1991) Increase of macrophage colony-stimulating factor and granulocyte-macrophage colony-stimulating factor receptors in the regenerating rat facial nucleus. J Neurosci Res 30: 682–686
9. Raivich G, Kreutzberg GW (1994) Pathophysiology of glial growth factor receptors. Glia 11: 129–146
10. Sendtner M, Kreutzberg GW, Thoenen H (1990) Ciliary neurotrophic factor prevents the degeneration of motoneurons after axotomy. Nature 345: 440–441

Correspondence: Georg W. Kreutzberg, M.D., Department of Neuromorphology, Max-Planck-Institute of Psychiatry, D-82152 Martinsried, Federal Republic of Germany.

Acta Neurochir (1996) [Suppl] 66: 107–113

Neuroprotective Efficacy of Microvascularly-Localized Versus Brain-Penetrating Antioxidants

E. D. Hall[1], P. K. Andrus[1], S. L. Smith[1], J. A. Oostveen[1], H. M. Scherch[1], B. S. Lutzke[1], T. J. Raub[2], G. A. Sawada[2], J. R. Palmer[3], L. S. Banitt[3], J. S. Tustin[3], K. L. Belonga[3], D. E. Ayer[3], and G. L. Bundy[3]

[1]CNS Diseases Research, [2]Drug Delivery Systems Research, and [3]Medicinal Chemistry Research, The Upjohn Company, Kalamazoo, MI, U.S.A.

Summary

The 21-aminosteroid (lazaroid) tirilazad mesylate has been demonstrated to be a potent inhibitor of lipid peroxidation and to reduce traumatic and ischemic damage in a number of experimental models. Currently, tirilazad is being actively investigated in Phase III clinical trials in head and spinal cord injury, ischemic stroke and subarachnoid hemorrhage. This compound acts in large part to protect the microvascular endothelium and consequently to maintain normal blood-brain barrier (BBB) permeability and cerebral blood flow autoregulatory mechanisms. However, due to its limited penetration into brain parenchyma, tirilazad has generally failed to affect delayed neuronal damage to the selectively vulnerable hippocampal CA1 and striatal regions. Recently, we have discovered a new group of antioxidant compounds, the pyrrolopyrimidines, which possess significantly improved ability to penetrate the BBB and gain direct access to neural tissue. Several compounds in the series, such as U-101033E, have demonstrated greater ability to protect the CA1 region in the gerbil transient forebrain ischemia model with a post-ischemic therapeutic window of at least four hours. In addition, U-101033E has been found to reduce infarct size in the mouse permanent middle cerebral artery occlusion model in contrast to tirilazad which is minimally effective. These results suggest that antioxidant compounds with improved brain parenchymal penetration are better able to limit certain types of ischemic brain damage compared to those which are localized in the cerebral microvasculature. On the other hand, microvascularly-localized agents like tirilazad appear to have better ability to limit BBB damage.

Keywords: Antioxidant; tirilazad; pyrrolopyrimidine; ischemia.

Introduction

There is now a significant amount of information that supports a role of oxygen radical-induced lipid peroxidation (LP) in the pathophysiology of acute central nervous system injury and ischemia [4,7,20]. The 21-aminosteroid (lazaroid) tirilazad mesylate has been demonstrated to be a potent inhibitor of LP that acts by a combination of chemical radical scavenging and membrane stabilization mechanisms. It has been shown to reduce traumatic and ischemic damage in a number of experimental models, and a correlation has been demonstrated in several instances between attenuation of oxygen radical levels and/or lipid peroxidation and the neuroprotective effect (see review by Hall *et al.*, [9]). Currently, tirilazad is being actively investigated in Phase III clinical trials in head and spinal cord injury, ischemic stroke and subarachnoid hemorrhage (SAH). Results from a multi-national European/Australian/New Zealand trial in SAH have demonstrated a highly significant reduction in 3 month mortality and improvement in the incidence of "Good" recovery (Glasgow Outcome Scale) in patients treated with tirilazad [11].

Tirilazad appears to act in large part on the microvascular endothelium [1,16] and consequently has been shown to protect the blood-brain barrier (BBB), to maintain cerebral or spinal cord blood flow autoregulatory mechanisms and/or to reduce delayed vasospasm in multiple models [9]. Therefore, its' ability to protect neural tissue from traumatic or ischemic insult in many models may be largely indirect. Indeed, tirilazad, most likely due to its limited penetration into brain parenchyma [19], has generally failed to affect delayed neuronal damage in the selectively vulnerable hippocampal CA1 and striatal regions [2,5,12,21], although it has some effect to protect cortical neurons [12,21]. Moreover, in models of permanent focal ischemia where microvascular effects may be less important than in temporary ischemia paradigms, the

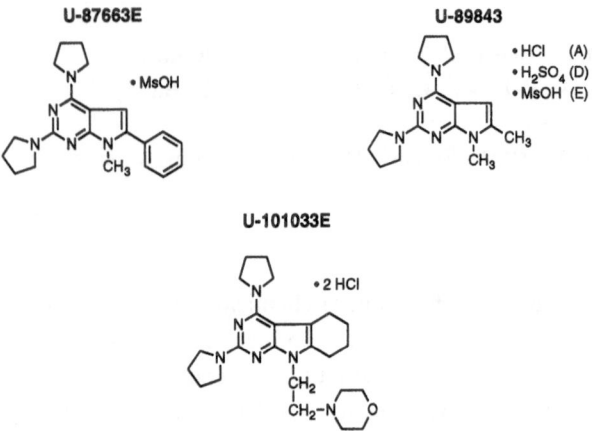

**Tirilazad Mesylate
(U-74006F)**

Pyrrolopyrimidines

U-87663E

U-89843

• MsOH

• HCl (A)
• H$_2$SO$_4$ (D)
• MsOH (E)

U-101033E

• 2 HCl

Fig. 1. Chemical structures of the 21-aminosteroid or lazaroid tirilazad mesylate (U-74006F) and selected pyrrolopyrimidines

which is inactive. On the other hand, microvascularly-localized agents like tirilazad appear to have better ability to limit BBB damage after experimental SAH.

Structural Comparison of Tirilazad Mesylate and the Pyrrolopyrimidines

Figure 1 displays the chemical structures of tirilazad mesylate and three of the pyrrolopyrimidines, U-87663E (the original prototype), U-89843D and U-101033E*. Tirilazad is a non-glucocorticoid 21-aminosteroid or "lazaroid". The primary chemical antioxidant portion of the molecule is the amino moiety bound to the 21 position of the steroid side chain. Physicochemical studies indicate that the highly lipophilic (i.e. hydrophobic) steroid moiety orients itself within the hydrophobic fatty acid core of the membrane and is largely responsible for the high affinity of the compound for cell (e.g. endothelium) membranes [9,19]. The more hydrophilic amino substitution exists closer to the surface in juxtaposition to the phosphate head groups of the phospholipids. In contrast, the pyrrolopyrimidines lack the highly lipophilic steroid moiety, while bearing some structural resemblance to the bispyrrolidinylpyrimidinyl piperazine 21-amino substitution of tirilazad. The lack of the steroid serves to lessen the high affinity for and retention in lipid bilayers (T. J. Raub and G. A. Sawada, unpublished results).

Comparison of Inhibition of Iron-Dependent Lipid Peroxidative Neuronal Injury

Table 1 shows the IC$_{50}$s and maximum % protection of cultured fetal mouse spinal neurons from iron (200 µM ferrous ammonium sulfate)-induced lipid peroxidative injury (system described in detail elsewhere, 8) by tirilazad in comparison to the pyrrolopyrimidines. Protection was measured in terms of preservation of amino acid uptake (i.e. uptake of 3H-amino isobutyric acid). As seen, the pyrrolopyrimidines are generally more potent and slightly more efficacious in this *in vitro* model. However, with both types of compounds, a correlation has been demonstrated between preservation of amino acid uptake and attenuation of iron-induced lipid peroxidation (data not shown).

compound's ability to affect infarct size, while demonstrated in some experiments [3,15], has been inconsistent [22].

Thus, we reasoned that lipid peroxidation-inhibiting (antioxidant) compounds with improved brain penetration might possess certain advantages over the microvascularly-localized tirilazad in certain CNS injury situations. Recently, we have discovered a new group of compounds, the pyrrolopyrimidines (Fig. 1), which are equal or better antioxidants than tirilazad, but with significantly improved ability to penetrate the BBB and gain direct access to neural tissue. Several compounds in the series, including U-101033E, have demonstrated greater ability than tirilazad to protect the CA1 region in the gerbil forebrain ischemia model with a post-ischemic therapeutic window of at least four hours. In addition, U-101033E has been found to reduce infarct size in the mouse permanent middle cerebral artery occlusion model in contrast to tirilazad

---

* The suffix letters indicate the salt form of the compound. Different salts were sometimes studied such as A = hydrochloride, D = sulfuric or E = methane sulfonate. Thus, suffix letters may vary below.

Table 1. *Comparison of the Potency and Efficacy of U-74006F and Selected Pyrrolopyrimidines with Regard: to Protection Against Lipid Peroxidative Impairment of Viability (i.e. Attenuation of ^{3}H-AIB Uptake) in Cultured Fetal Mouse Spinal Neurons by 200 μM Ferrous Ammonium Sulfate Application for 40 min.* See ref. [8] for detailed methods

Compound	$IC_{50}(\mu M)$	Maximum protection (% of control)
U-74006F[a]	5.4–25.4	57.2–60.5
U-87663E	1.3	100
U-89843D	2.3	100
U-101033E	1.1	75.8

[a] Range from multiple trials.

Table 2. *Comparison of the Brain Levels of U-74006F and Selected Pyrrolopyrimidines at 5 or 60 min. After a 23 $\mu M/kg$ (Approx. 10 mg/kg) i.v. Bolus.* Mice were perfused with saline at sacrifice to eliminate contribution of blood to the measured levels

Compound	5 min. (nmoles/g)	60 min. (nmoles/g)
U-74006F (N = 4)	3.8±1.6	2.9±0.5
U-87663E (N = 2)	17.7±0.8[a]	6.7±2.6[a]
U-89843E (N = 3)	12.1±0.8[a]	1.6±0.1
U-101033E (N=3)	10.7±2.0[a]	0.6±0.3

Values = means ± standard error.

[a] $p < 0.05$ compared to U-74006F at the same time point.

Comparison of Brain Uptake

The comparative brain uptake of tirilazad and the pyrrolopyrimidines in mice after i.v. administration of molar equivalent doses (approx. 10 mg/kg) is shown in Table 2. Each of the pyrrolopyrimidines produced significantly higher brain levels at 5 min. after injection compared to tirilazad. For U-87663E, U-89843E and U-101033E, the brain levels at the initial time point were 4.7, 3.2 and 2.8x higher, respectively, than the levels of tirilazad. In the case of U-87663E, the brain levels were still 2.3x higher than tirilazad at 60 min. post-injection. It is also clear that the brain levels seem to fall off faster for the pyrrolopyrimidines as a further reflection of their greater membrane permeabilities (T. J. Raub and G. A. Sawada, unpublished results). In other words, they diffuse into the brain, and, in the absence of repeated dosing, they may diffuse out of the brain more quickly than tirilazad.

We have also measured the brain uptake of tirilazad and selected pyrrolopyrimidines in rats in terms of first pass extraction of radiolabeled compounds after intra-carotid injection (13). Using this technique, only 6 % of tirilazad is extracted which is not much better than the extraction of sucrose or thiourea to which the BBB is essentially nonpermeable (Fig. 2). In contrast, 83 % of

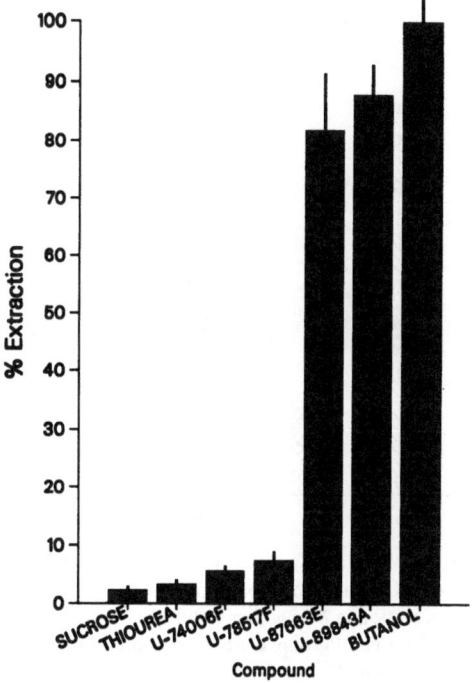

Fig. 2. Comparison of the brain uptake (first pass extraction) of U-74006F compared to the pyrrolopyrimidines U-87663E and U-89843A after an intra-carotid injection (Oldendorf method [13]). Values = means ± standard deviation for 3–27 animals/compound. Also compared are sucrose and thiourea which essentially do not penetrate the blood-brain barrier and butanol which is freely permeable. U-78517F is a 2-methylaminochroman antioxidant which possesses the same amino group as U-74006F, but with the ring structure of vitamin E in place of the steroid moiety [8]

the pyrrolopyrimidine U-87663E is extracted on the first pass through the cerebral circulation or 88 % of U-89843E. These extraction values are very near that measured for the freely BBB-diffusible butanol. Thus, these pyrrolopyrimidines appear to permeate the BBB quite readily. Indeed, other work has taken advantage of the intrinsic fluorescent properties of U-87663E to demonstrate unequivocally by fluorescence microscopy that this prototype pyrrolopyrimidine efficiently penetrates the BBB in mice after i.v. dosing. Additionally, it concentrates in brain parenchyma and not in the cerebrovascular endothelium. Confocal laser microscopy has also been used to show that U-87663E readily gains access to the intracellular space of cultured cells (T. J. Raub and G. A. Sawada, unpublished results).

Comparison of Neuronal Protection in Transient Forebrain Ischemia

As noted above, tirilazad mesylate has generally demonstrated only weak ability to attenuate selective hippocampal CA1 vulnerability in models of transient

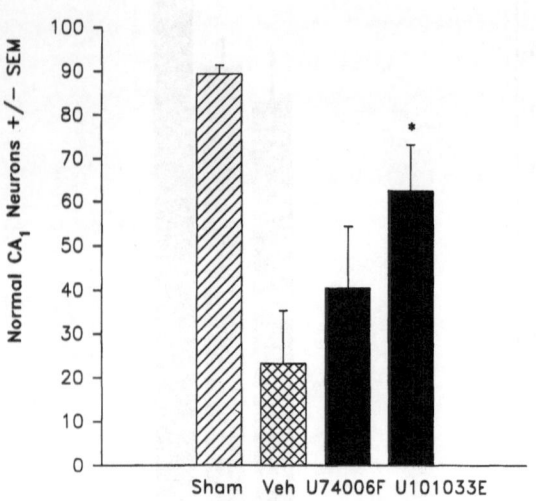

Fig. 3. Comparison of the ability of U-74006F and U-101033E to salvage hippocampal CA1 neurons at 5 days after a 5 min episode of bilateral carotid occlusion in Mongolian gerbils. Values = means ± standard error for 10 animals/group. Gerbils were dosed with 30 mg/kg per os pre-ischemia plus 2 hrs after reperfusion and once daily on days 2,3 and 4. Asterisk indicates p < 0.05 vs. the vehicle treated group

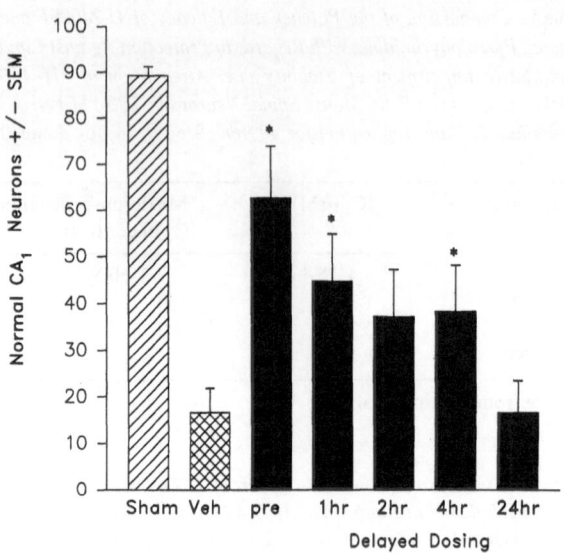

Fig. 5. Therapeutic window of U-101033E to salvage hippocampal CA1 neurons at 5 days after a 5 min episode of bilateral carotid occlusion in Mongolians gerbils. Values = means ± standard error for 10 animals/group. Gerbils were dosed beginning at each time point with 30 mg/kg per os plus 2 hrs later and on the subsequent days. Asterisk indicates p < 0.05 vs. the vehicle treated group

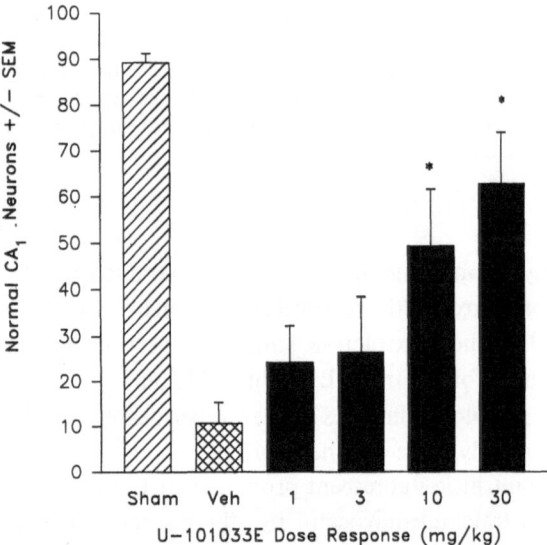

Fig. 4. Dose-response of U-101033E to salvage hippocampal CA1 neurons at 5 days after a 5 min episode of bilateral carotid occlusion in Mongolian gerbils. Values = means ± standard error for 10 animals/group. Gerbils were dosed with each dose level per os pre-ischemia plus 2 hrs after reperfusion and once daily on days 2,3 and 4. Asterisk indicates p < 0.05 vs. the vehicle treated group.

oral pretreatment with the more BBB-permeable pyrrolopyrimidine U-101033E produces twice as much CA1 protection compared to tirilazad, reaching a level that is significantly greater than the CA1 preservation observed in vehicle treated animals. Figure 4 shows the dose-response curve for the ability of U-101033E pretreatment to protect the CA1 region. As seen, dose levels of 10 or 30 mg/kg are significantly effective, whereas doses as low as 1 and 3 mg/kg appear to have only some effect.

Therapeutic Window in Transient Forebrain Ischemia

Figure 5 displays the therapeutic window for the efficacy of U-101033E with regard to CA1 protection in the gerbil 5 min. forebrain ischemia model. As can be seen in this graph, the initiation of dosing 30 min. prior to ischemia (plus repeated post-ischemic dosing) is the most effective. However, a delay in dosing to 4 hr. after reperfusion still provides a statistically significant neuroprotective effect.

Comparison of Infarct Reduction in Permanent Focal Ischemia

The neuroprotective efficacy of tirilazad has also been compared to selected pyrrolopyrimidines in the context of focal ischemic models. For instance, both

forebrain ischemia [2,5,12,15] perhaps due to its limited BBB penetration in the context of models where BBB permeability is minimally compromised. Similarly Fig. 3 shows that preischemic oral treatment with tirilazad (plus additional post-ischemic dosing) has only a small effect on CA1 neuronal preservation in the gerbil 5 min. forebrain ischemia model. In contrast,

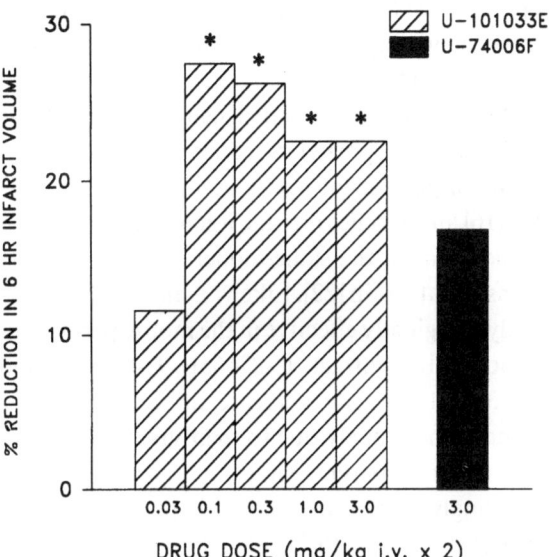

Fig. 6. Comparison of the ability of U-74006F and U-101033E to reduce the 6 hr infarct size in male CF-1 mice subjected to permanent unilateral occlusion of the right middle cerebral artery. The mice received one dose of compound i.v. at 5 min post-occlusion and a second dose at 1 hr Values = percent reduction in mean infarct volume (summation of 8 cross sections through the infarct) compared to a vehicle-treated group of animals. N = 8–13 animals/group. Asterisk indicates p < 0.05 vs. paired vehicle

Table 3. *Comparison of the Effects of U-74006F and Selected Pyrrolopyrimidines on the Neuronal Damage Index[a] and Neutrophil Influx[b] in the Hippocampal CA1 Region of Male Mongolian Gerbils Following 3 hrs of Unilateral Carotid Occlusion-Induced Hemispheric Ischemia and 12 hrs of Reperfusion.* Gerbils received a 10 mg/kg i.p. dose of either vehicle or test compound at 10 min. prior to carotid occlusion and a second dose at the time of reperfusion

Compound	Neuronal damage index	Neutrophil influx
Vehicle (N = 16)	2.6±0.2	108.9±12.4
U-74006F (N = 6)	1.7±0.4[c]	25.9±10.8[c]
U-87663E (N = 7)	1.6±0.5[c]	55.3±12.6[c]
U-89843A (N = 6)	1.7±0.3[c]	51.8±7.6[c]

Values = means ± standard error.
[a] 0 = no damage, 1 = < 25 % loss of CA1 neurons, 2 = 25–50 % loss, 3 = 50–75 % loss and 4 = > 75 % loss.
[b] Neutrophils/mm^2.
[c] p < 0.05 compared to vehicle.

have been examined for their abilities to limit brain infarct volume in mice subjected to permanent middle cerebral artery (MCA) occlusion. Mice received two doses of test compound, one at 5 min. and a second at 60 min. post-occlusion. As observed in Fig. 6, U-101033E potently reduces infarct volume by as much as 27 % at a dose of only 0.1 mg/kg (x2) i.v. In contrast, tirilazad at a dose of 3 mg/kg (x2) i.v., while having some effect, was not significant. In addition, U-87663E and U-89843A have also been found to be significantly

effective in the mouse permanent MCA occlusion model, while tirilazad (0.1–3.0 mg/kg x2) has shown only non-significant trends toward infarct reduction in repeated experiments (data not shown).

Comparison of Neuronal Protection and Attenuation of Neutrophil Invasion in Temporary Focal Ischemia

Tirilazad and certain pyrrolopyrimidines have also been examined in a model of temporary focal ischemia, the gerbil 3 hr unilateral carotid occlusion paradigm [10], concerning their relative efficacy in decreasing 12 hr post-reperfusion neuronal damage and neutrophil influx. Table 3 indicates that in this model of temporary focal ischemia,the abilities of tirilazad and either of the pyrrolopyrimidines U-87663E or U-89843A to reduce post-ischemic necrosis to the hippocampal CA1 region are approximately equal. All three decrease neuronal damage by 35–38 %. Similarly, all three compounds tested significantly limit neutrophil influx, although the effect of tirilazad is more impressive.

Comparison of Preservation of BBB Integrity in Acute SAH

The microvascularly-localized tirilazad and the brain-penetrable pyrrolopyrimidine U-87663E have also been compared in terms of their abilities to prevent BBB disruption in rats during the first 3 hrs after SAH when equivalently dosed (1 mg/kg i.v. x2). Figure 7 illustrates that tirilazad outperforms U-87663E in this context. While the former produced a highly significant > 50 % attenuation of post-SAH Evans's blue extravasation, U-87663E produced only a modest effect.

Discussion

The present report describes a novel group of potent inhibitors of iron-dependent lipid peroxidation in the neural tissue, the pyrrolopyrimidines (e.g. U-87663E, U-89843D, U-101033E). They possess improved BBB permeability and brain parenchymal penetration compared to tirilazad which remains largely localized in the cerebral microvascular endothelium [9,19]. As a result, compounds like U-101033E are much better than tirilazad at attenuating selective neuronal damage in highly vulnerable regions, such as the CA1 area of the hippocampus following a transient episode of forebrain ischemia. Other supposedly brain penetrable an-

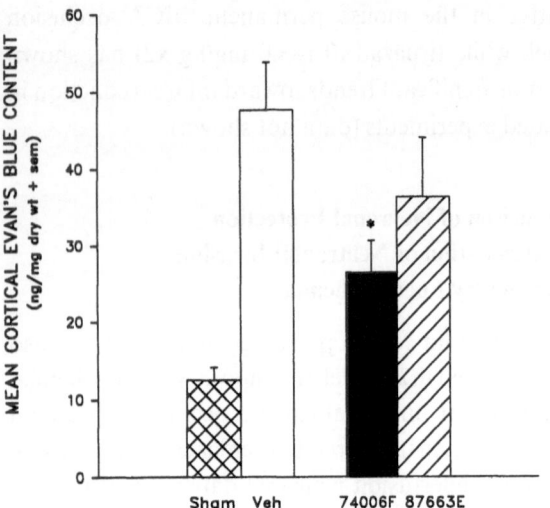

Fig. 7. Comparison of the ability of U-74006F and the pyrrolopyrimidine U-87663E to reduce blood-brain barrier permeability (Evans' blue extravasation) in male Sprague-Dawley rats at 3 hrs after SAH. The animals received a 1 mg/kg i.v. dose of either compound (or an equivalent volume of vehicle) at 15 min before and again at 2 hrs after SAH. Brains were dried and the Evan's blue extracted with formamide and measured fluorometrically. Values = means ± standard error. N = 8–14 animals/group. Asterisk indicates p < 0.05 vs. vehicle

tioxidants have been previously described with neuroprotective efficacy in transient forebrain ischemia models in either gerbils or rats including LY-178002 [6], N-tert-butyl-α-phenynitrone (PBN; [17]) and dimethylthiourea (DMTU; [14]). However, much higher doses of all of these compounds appear to be required to achieve neuroprotection as compared to U-101033E which is significantly effective at oral dose levels as low as 10 mg/kg. This suggests that these earlier-described and studied compounds either may not be as brain-penetrable as thought or perhaps they are not as effective at attenuating oxygen radical-induced, iron-catalyzed lipid peroxidation as the pyrrolopyrimidines appear to be. In addition, U-101033E has been shown to have at least a 4 hr post-ischemic therapeutic window. In contrast, PBN's ability to protect CA1 neurons in the identical gerbil forebrain ischemia model is lost by 2 hrs after reperfusion [17]. Nevertheless, further study of U-101033E and other pyrrolopyrimidines is neccessary before an exact assessment of their neuroprotective activity in comparison to the earlier described compounds can be firmly established.

The pyrrolopyrimidines similarly outperform tirilazad, as well as another antioxidant dihydrolipoate [18], with regard to the ability to reduce early infarct growth in the mouse permanent MCA occlusion model. In the face of permanent vascular occlusion, a successful

neuroprotective compound must intuitively be able to penetrate the underperfused ischemic penumbra zone in order to be optimally effective in salvaging the still viable, but potentially doomed neural tissue. While tirilazad has been reported to reduce infarct volume in the setting of permanent MCA occlusion in Sprague-Dawley [16] and Fischer [3] rats, it has not been shown to be efficacious in the same model in spontaneously hypertensive rat strain [22]. Likewise, tirilazad appears to be only marginally effective in the mouse permanent MCA model. In contrast, U-87663E, U-89843D and U-101033E all potently decrease infarct size in the same model, most likely due to their greater access to the ischemic brain parenchyma.

Interestingly, the brain-penetrable pyrrolopyrimidines are not superior to tirilazad in the context of either temporary (but prolonged) focal ischemia (gerbil 3 hr unilateral carotid occlusion) or in SAH-induced BBB damage. In the setting of temporary focal ischemia, tirilazad and U-87663E and U-89843A are equally neuroprotective. Two explanations are feasible. First of all, post-reperfusion BBB damage in this model may serve to enhance parenchymal access of tirilazad such that it is able to reach the neurophil. Secondly, BBB damage, and its potential attenuation by the micro-vascularly-localized tirilazad, may be equally important with parenchymal neuronal injury mechanisms which would be most efficiently countered by the brain-penetrable pyrrolopyrimidines. On the other hand, in the latter setting of SAH-induced BBB damage and the consequent pathological protein permeability, it is apparent that the microvascularly-localized tirilazad has a clear advantage. Thus, within the overall spectrum of traumatic, ischemic and hemorrhagic CNS insults, it would seem that antioxidant compounds that localize in brain microvasculature or that penetrate the brain parenchyma will both have specific, and no doubt complimentary, therapeutic roles to play.

References

1. Audus KL, Guillot FL, Braughler JM (1991) Evidence for 21-aminosteroid association with the hydrophobic domains of brain microvessel endothelial cells. Free Rad Biol Med 11: 361–371
2. Beck T, Bielenberg GW (1990) Failure of the lipid peroxidation inhibitor U-74006F to improve neurologic outcome after transient forebrain ischemia in the rat. Brain Res 532: 336–338
3. Beck T, Bielenberg GW (1991) The effects of two 21-aminosteroids on overt infarct size 48 hours after middle cerebral artery occlusion in the rat. Brain Res 560: 159–162
4. Braughler JM, Hall ED (1989) Central nervous system trauma and stroke: I. Biochemical considerations for oxygen radical formation and lipid peroxidation. Free Red Biol Med 6: 289–301

5. Buchan AM, Bruederlin B, Heinicke E, Li H (1992) Failure of the lipid peroxidation inhibitor U-74006F to prevent postischemic selective neuronal injury. J Cereb Blood Flow Metab 12: 250–256

6. Clemens JA, Ho PPK, Panetta JA (1991) LY178002 reduces rat brain damage after transient global forebrain ischemia. Stroke 22: 1048–1052

7. Hall ED, Braughler JM (1989) Central nervous system trauma and stroke: II. Physiological and pharmacological evidence for the involvement of oxygen radicals and lipid peroxidation. Free Rad Biol Med 6: 303–313

8. Hall ED, Braughler JM, Yonkers PA, Smith SL, Linseman KL, Means ED, Scherch HM, Jacobsen EJ, Lahti RA (1991) U-78517F: a potent inhibitor of lipid peroxidation with activity in experimental brain injury and ischemia. J Pharmacol Exp Ther 258: 688–694

9. Hall ED, McCall JM, Means ED (1994) Therapeutic potential of the lazaroids (21-aminosteroids) in central nervous system trauma, ischemia and subarachnoid hemorrhage. Adv Pharmacol 28: 221–268

10. Hall ED, Pazara KE, Braughler JM (1988) 21-Aminosteroid lipid peroxidation inhibitor U-74006F protects against cerebral ischemia in gerbils. Stroke 19: 997–1002

11. Kassell NF, Haley EC, Apperson-Hansen C, Alves WM Dovsch NW, Fabinyï G, Matheson J, Reilly P, Siu, K et al. (1996) Randomized, double-blind, vehicle-controlled trial of tirilazad mesylate in patients with aneurysmal subarachnoid hemorrhage: A Cooperative Study in Europe, Australia, and NewZealand. J Neurosurg 84: 221–228

12. Lesiuk HJ, Sutherland GR, Peeling J, Wilkins D, McTavish J, Saunders JK (1991) Effect of U-74006F on forebrain ischemia in rats. Stroke 22: 896–901

13. Oldendorf WH (1981) Clearance of radiolabelled substances by brain after arterial injection using a diffusible internal standard. Res Meth Neurochem 5: 91–112

14. Pahlmark K, Folbergrova J, Smith M-L, Siesjö BK (1993) Effects of dimethylthiourea on selective neuronal vulnerability in forebrain ischemia in rats. Stroke 24: 731–737

15. Pahlmark K, Smith ML, Siesjö BK (1991) Failure of U-74006F to ameliorate neuronal damage due to transient ischemia or hypoglycemia. J Cereb Blood Flow Metab 11 [Suppl 2]: S138

16. Park CK, Hall ED (1994) Dose-response analysis of the 21-aminosteroid tirilazad mesylate (U-74006F) upon neurological outcome and ischemic brain damage in permanent focal cerebral ischemia. Brain Res 645: 157–163

17. Phillis JW, Clough-Helfman C (1990) Protection from cerebral ischemic injury in gerbils with the spin trap agent N-tert-butyl-α-phenylnitrone (PBN). Neurosci Lett 116: 315–319

18. Prehn, JHM, Karkoutly C, Nuglisch J, Peruche B, Krieglstein J (1992) Dihydrolipoate reduces neuronal injury after cerebral ischemia. J Cereb Blood Flow Metab 12: 78–87

19. Raub TJ, Barsuhn CL, Williams LR, Decker DE, Sawada GA, Ho NFH (1993) Use of a biophysical-kinetic model to understand the roles of protein binding and membrane partitioning on passive diffusion of highly lipophilic molecules across cellular barriers. J Drug Targeting 1: 269–286

20. Siesjö BK, Agardh C-D, Bengtsson F (1989) Free radicals and brain damage. Cerebrovasc Brain Metab Rev 1: 165–211

21. Sutherland G, Haas N, Peeling J (1993) Ischemic neocortical protection with U-74006F- a dose-response curve. Neurosci Letts 149: 123–125

22. Xue D, Slivka A, Buchan AM (1992) Tirilazad reduces cortical infarction after transient, but not permanent focal cerebral ischemia in rats. Stroke 23: 894–899

23. Zuccarello M, Anderson DK (1989) Protective effect of a 21-aminosteroid on the blood-brain barrier following subarachnoid hemorrhage in rats. Stroke 20: 367–371

Correspondence: Edward D. Hall, Ph.D., CNS Diseases Research, The Upjohn Company, Kalamazoo, MI 49001, U.S.A.

Acta Neurochir (1996) [Suppl] 66: 114–117

"Small Volume Resuscitation" as Treatment of Cerebral Blood Flow Disturbances and Increased ICP in Trauma and Ischemia

O. Kempski[1], **C. Obert**[1], **T. Mainka**[1], **A. Heimann**[1], and **U. Strecker**[2]

[1]Institute for Neurosurgical Pathophysiology and [2]Department of Anesthesia, Johannes Gutenberg-Universität Mainz, Mainz, Federal Republic of Germany

Summary

"Small volume resuscitation" (SVR) is a promising concept for the treatment of shock and trauma patients. SVR utilizes the fast infusion of a small volume of hypertonic saline to mobilize intraendothelial and parenchymal water to expand and restitute intravascular volume. Therefore it seems warranted to also consider SVR for the treatment of disturbances of the cerebral circulation and of increased intracranial pressure (ICP). The current study uses a rabbit model of global cerebral ischemia combined with mild hemorrhage to test SVR. Somatosensory evoked potentials (SEPs) serve as a short-term outcome parameter. The data demonstrate a beneficial effect on ICP, a better reperfusion, and an improved recovery of SEPs after treatment with hypertonic/hyperoncotic saline/hydroxyethylstarch as compared to volume replacement with starch or blood.

Keywords: Small volume resuscitation; cerebral ischemia; head injury; intracranial pressure.

Introduction

"Small volume resuscitation" (SVR) has been introduced as a promising concept for the treatment of shock and trauma patients [4,12,16,23]. SVR utilizes the fast infusion of a small volume of hypertonic saline to mobilize intraendothelial and parenchymal water to expand and restitute intravascular volume. Thereby, as a beneficial effect, hemodilution together with the hypertonicity-induced shrinking of the endothelial/perivascular volume serve to significantly improve the microcirculation, and to flush out potentially harmful metabolites (Fig. 1). Hypertonicity is achieved by the infusion of 7.5 % NaCl. In order to maintain the volume effect, a hyperoncotic component is generally added, e.g., dextran or hydroxyethyl starch [12]. For shock treatment, only 10 % of the actual volume loss has to be infused in order to stabilize the circulation. In addition to the volume recovery, an increase in myocardial contractility and a decrease of peripheral resistance result in an increase of cardiac output, responsible for the better perfusion. Hypertonic NaCl-solutions are superior in restoring the membrane potential and shock related cellular electrolyte imbalance [14]. Particularly brain, heart and liver profit from these effects (cf. [22]).

SVR has been originally designed for prehospital emergency treatment of hemorrhagic shock. The progress of the pathophysiological understanding of the mechanisms involved suggests other applications: after global cerebral ischemia in particular, reperfusion injury of the brain apparently evolves along pathways similar to those seen in the peripheral circulation in shock. A multitude of mediator cascades-free radicals,

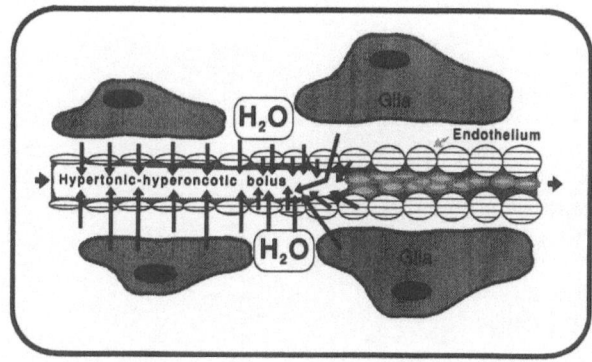

Fig. 1. Schematic drawing of the effect of hypertonic/hyperoncotic solutions on the cerebral microcirculation. The osmotic gradient draws fluid from the extravascular space - endothelium, extracellular compartment and surrounding glial cells. This improves the rheology, facilitates reperfusion after ischemia, reduces brain edema if present, and in addition leads to a fast washout of toxic metabolites and eventually adhering neutrophils

eicosanoids, activated leukocytes just to name a few - are apparently involved. Therefore, a fast and efficient recirculation has been advertised as being far superior to early but trickling reflow [8]. Attempts to treat stroke patients with hemodilution underline this concept [3,5,6,11,15,19]. In addition the brain is particularly sensitive to ischemia and reacts with perivascular swelling of glial cells to even short ischemic episodes. Taken together, the current pathophysiological understanding of global cerebral ischemia suggest that SVR might have positive effects during the early reperfusion phase.

The current study was hence designed to elucidate two major questions: a) does SVR have negative effects on the brain in a model of global cerebral ischemia combined with blood loss, or b) does outcome from ischemia rather benefit from SVR? Negative effects might include rises of intracranial pressure (ICP) resulting from damage to the blood-brain barrier. Vasogenic edema formation via a defective barrier has been observed in cases with a significantly enhanced postischemic hyperperfusion [13,20]. Conversely, our own data have shown a better outcome from global ischemia to be associated with enhanced hyperperfusion in the gerbil model [10].

To address these questions cerebral macro- and microcirculation, ICP, and somatosensory evoked potentials (SEP) as an outcome parameter were monitored in a rabbit model of global cerebral ischemia with preischemic blood loss. Together with recirculation either hypertonic-hyperoncotic hydroxyethyl starch (HHES), normotonic hydroxyethyl starch (HES) or the shed blood (BLOOD) were reinfused.

Methods

Female New Zealand rabbits (5.0 ± 0.5 kg; n = 10 per group) received 28 mg/kg bw (body weight) thiopental and 0.4 mg/kg bw pancuronium bromide for induction of anesthesia and oral intubation. Anesthesia was maintained with 1 % alpha-chloralose (50 mg/kg bw; 40 ml/h). For controlled ventilation with a 33 % oxygen/air mixture and 0.3 mg/kg/h pancuronium bromide, an end-expiratory pCO_2 of 34 mmHg was aimed at. Body temperature was maintained at 39 °C. Per hour the animals received 25 IU/kg bw heparin and 6ml/kg bw ringer lactate solution.

A fiberoptic oxygen saturation catheter was introduced into the internal jugular vein. Additional catheters to measure blood pressures and to sample blood were advanced into the aorta via the femoral artery, and into the inferior vena cava via the femoral vein. The right cerebral ventricle was cannulated for the continuous registration of intracranial pressure. Following a mediastinotomy ligatures were lightly placed around all vessels leaving the aortic arch, and guided through the chest wall via plastic tubing for later ischemia induction. An electromagnetic flow probe (Hellige Recomed, Freiburg/Germany) was placed around the left common carotid

artery. Craniectomies were drilled to position transcranial Doppler (8 MHz, EME TC 2000, Überlingen/Germany), and laser Doppler (TSI Laserflo, St. Paul MN, USA) probes. After a control period and the withdrawal of blood (10 ml/kg bw) the ligatures were simultaneously occluded for 15 min. After onset of reperfusion the infusions were initiated according to the following protocol:

1. HHES: 7.5 % NaCl + 10 % hydroxyethyl starch 200,000/ 0.5, 5 ml/kg bw;
2. HES: 10 % hydroxyethyl starch 200,000/0.5, 10 ml/kg bw;
3. BLOOD: Reinfusion of the shed blood (+ 1500 IU heparin), 10 ml/kg bw.

Before, during, and for 5 h after ischemia the following parameters were recorded: extracranial blood flow by the electromagnetic flow probe around the carotid artery, intracranial macrocirculatory input by transcranial ultrasound Doppler of basilar artery velocity, parietal microcirculation (ICBF) by laser Doppler, internal jugular vein O_2- saturation, ICP, mean arterial pressure (MAP), central venous pressure, brain and body core temperature, SEPs, end-expiratory pCO_2, blood gases, electrolytes, and glucose.

Results

Recirculation was accompanied in all animals by a temporary increase of ICP from an initial average 3 to 8–9 mmHg (Table 1). In the HHES group, ICP normalized to 3.3 mmHg within 20 min, whereas a rather incomplete normalization required as long as 100–120 min with HES or BLOOD (p < 0.01). The fast ICP normalization with HHES was particularly impressive since the postischemic hyperperfusion was significantly more pronounced and commenced earlier with HHES than with both other treatments: After an

Table 1. *Intracranial Pressure (ICP) Before, During and After 15 min of Global Cerebral Ischemia in Rabbits*

[mmHg]	Control	Ischemia	2 min reperf.	5 min reperf.	10 min reperf.	15 min reperf.
HHES	3.2±0.5	0.73±0.25	6.9±1.28	8.3±0.27	5.8±0.68	5.0±0.84
BLOOD	3.0±0.48	1.35±0.34	4.9±1.57	8.3±2.23	8.7±2.02	7.7±0.73
HES	2.9±0.42	1.19±0.42	6.9±0.85	9.1±1.28	8.8±1.36	7.1±0.91

[mmHg]	20 min reperf.	25 min reperf.	30 min reperf.	45 min reperf.	60 min reperf.	240 min reperf.
HHES	3.3±0.73	1.9±0.39	2.3±0.55	2.2±0.82	1.9±0.3	3.8±0.5
BLOOD	6.3±0.86	5.9±0.91	6.0±0.69	5.3±0.56	4.2±0.58	3.3±0.75
HES	5.9±0.59	4.9±0.62	4.9±1.07	4.3±0.52	4.1±0.47	4.7±0.71

Immediately before ischemia 10 ml/kg bw blood were withdrawn. Within the initial 10 min of reperfusion each animal received either its own shed blood (BLOOD n = 10), hydroxyethyl starch (HES n = 10), or 7.5 % NaCl + hydroxyethyl starch (HHES n = 10). Postischemic hyperperfusion was accompanied in all animals by a temporary increase of ICP to 8–9 mmHg. In spite of the increased cerebral perfusion in the HHES group, ICP normalized to 3 mmHg within 20 min, whereas normalization was not achieved with HES or BLOOD. ICP in the HHES group remained significantly lower than in both other groups during the first 60 min of reperfusion (p < 0.01).

ischemic reduction of microcirculatory flow to $10 \pm 9\%$ of control, all animals showed considerable postischemic hyperperfusion, which, however, had quite a variable onset. This was reflected in changes of the basilar artery velocity but less in carotid flow. The normalization of ICBF, and the subsequent hyperperfusion occurred faster in the HHES group (after 5–7 min), than with HES (12–14 min). This difference of reperfusion was reflected in the significantly ($p < 0.01$) more elevated basilar artery velocity (30–50% for 25–30 min) as compared to HES, and to the even lower ($p < 0.01$) velocities in the BLOOD group. The protracted onset of reperfusion in the HES group went along with a slower recovery of the jugular vein O_2-saturation.

There was no statistical difference of arterial blood pressure between the three groups. Hematocrit was nearly unchanged as compared to the initial control level after reinfusion of the shed blood (BLOOD). In contrast to this, hematocrit was reduced from 32.9 to 21.4% in the HHES group, and from 30.1 to 22.5% in the HES group. Five hours of reperfusion were survived by all animals in the HHES and BLOOD group, while two out of ten rabbits died in the HES group. The recovery of the SEP amplitudes correlated with the better reperfusion after HHES treatment: after an initial loss of the SEP responses during ischemia, they returned to 30–40% of the initial amplitude in all three groups within 1 h of reperfusion. Then, however, the recovery continued only in the HHES group (to 81% after 5 h, $p < 0.01$), whereas in the two other groups there was no significant further improvement.

Discussion

SVR did not worsen the outcome during 5-h recovery from a 15-min period of global cerebral ischemia. On the contrary, outcome – as assessed by SEP-recovery but also survival – was significantly better than in both other experimental groups. Of particular clinical significance is the fast normalization of intracranial pressure (Table 1). The initial increase after ischemia seen in all groups, is due to cytotoxic edema formation on the one hand, and the increase of blood volume due to hyperperfusion on the other hand. Hypertonic HES most likely reduces ICP by osmotic shrinkage of swollen endothelial and glial cells, thereby reducing cytotoxic edema and facilitating reperfusion. ICP reduction may assume critical significance in patients with head injury with an already reduced intracranial compliance. In such patients any additional ICP increase, e.g., after an ischemic episode in intensive care, can have fatal consequences. Therefore all measures facilitating ICP recovery are of utmost importance. Experiments on hemorrhagic hypotension in dogs are in line with the data presented here: with a reduced cerebral perfusion pressure and elevated ICP, SVR reduced ICP and increased CBF as compared to isotonic saline infusion [18]. In a similar study Gunnar et al. could show [7] that in the presence of an intracranial mass lesion in beagle dogs resuscitation with hypertonic (3%) saline was accompanied by lower ICP values and less cerebral edema than were isotonic saline or colloid resuscitation. Recent data by Berger et al. [1] even support the conclusion that hypertonic/hyperoncotic saline/dextran reduces as efficiently as mannitol intracranial pressure raised by a cerebral lesion and a space-occupying mass. Both treatments might even be applied in alternation since the use of both is limited because of the possible rebound effect of mannitol and the effect of SVR on plasma sodium, and since the mechanisms of action are probably different – mannitol reduced water content of the traumatized hemisphere more effectively, whereas hypertonic/hyperoncotic saline/dextran was more effective in the contralateral uninjured side [2].

The fast normalization of ICP in the current study is also an indication that blood-brain barrier breakdown was not enhanced by the tremendous increase of postischemic hyperperfusion as described for focal ischemia with large increase of postischemic hyperperfusion [13,20]: had this been the case, the exudation of vasogenic edema would have resulted in a further increase of ICP, and not an accelerated normalization. SVR has recently been shown to reduce macromolecular leakage after muscle ischemia [17]. Apparently blood-brain barrier disruption during the hyperperfusion phase occurs, if hypervolemic hemodilution is combined with systemic hypertension [3]. A decisive advantage of SVR is the absence of blood pressure surges or longer periods of elevated arterial pressure, making SVR an excellent treatment candidate for stroke studies. A beneficial side effect for the treatment of stroke might be the increase of cardiac output [9] after SVR [21]. In rabbits Schürer et al. [21] saw increases of cardiac output (CO) directly after infusion of hypertonic/hyperoncotic saline/dextran from 81 to 137 ml/kg/min, and after 60 min CO was still increased. After hemorrhagic hypotension on the other hand CO was immediately normalized by the treatment.

In the current study the fast and almost complete recovery of the somatosensory evoked potential response after HHES is an indication that in this group the earlier and better washout of toxic metabolites together with the fast supply of the brain with oxygen had a therapeutic effect on reperfusion injury. "Small volume resuscitation" in its initial phase causes hypervolemic hemodilution and, in addition, facilitates reperfusion of the microcirculation by reducing local perivascular edema, and, thereby "no reflow" effects. The current data using HES without hypertonic saline did not exhibit any beneficial effect either on outcome or reperfusion as compared to reinfusion of the shed blood. This is in accordance with the observed inefficiency of isovolemic hemodilution in stroke [15].

Certainly the current data on *global* cerebral ischemia do not suffice to recommend SVR for the treatment of focal ischemic lesions. Results obtained by hypervolemic hemodilution trials, however, together with the present data are encouraging to perform studies employing SVR in models of focal ischemia.

References

1. Berger S, Schürer L, Härtl R, Messmer K, Baethmann A (1995) Reduction of posttraumatic hypertension by hypertonic/hyperoncotic saline/dextran and mannitol. Neurosurgery 37: 98–105
2. Berger S, Schürer L, Härtl R, Deisböck T, Dautermann C, Murr R, Messmer K, Baethmann A (1994) 7.2 % NaCl/10 % dextran 60 versus 20 % mannitol for treatment of intracranial hypertension. Acta Neurochir (Wien) [Suppl] 60: 494–498
3. Cole DJ, Drummond JC, Matsumura JS, Marcantonio S, Chi-Lum BI (1990) Hypervolemic hemodilution and hypertension during temporary middle cerebral artery occlusion in rats: the effect in blood-brain barrier permeability. Can J Neurol Sci 17: 372–377
4. DeFelipe J, Timoner J, Velasco IT, Lopes OU, Rocha-e-Silva M (1980) Treatment of refractory hypovolaemic shock by 7.5 % sodium chloride injections. Lancet 2: 1002–1004
5. Goslinga H, Eijzenbach V, Heuvelmans JHA, van der Laan de Vries E, Melis VMJ, Schmid-Schönbein H, Bezemer PD (1992) Custom-tailored hemodilution with albumin and cristalloids in acute ischemic stroke. Stroke 23: 181–188
6. Gottstein U (1981) Normovolemic and hypervolemic hemodilution in cerebrovascular ischemia. Bibliothca Haematol 47: 127–138
7. Gunnar W, Jonasson O, Merlotti G, Stone J, Barrett J (1988) Head injury and hemorrhagic shock: studies of the blood brain barrier and intracranial pressure after resuscitation with normal saline solution, 3 % saline solution, and dextran-40. Surgery 103: 398–407
8. Hossmann KA (1988) Resuscitation potentials after prolonged global cerebral ischemia in cats. Crit Care Med 16: 964–971
9. Keller TS, McGillicuddy JE, LaBond VA, Kindt GW (1985) Modification of focal cerebral ischemia by cardiac output augmentation. J Surg Res 39: 420–432
10. Kempski O, Stummer W, Bauer C, Schürer L, Baethmann A (1992) Postischemic hyperperfusion and outcome: an experimental study. In: Schmiedek P, Einhäupl K, Kirsch CM (eds) Stimulated cerebral bloodflow. Springer, Berlin Heidelberg New York Tokyo, pp 55–58
11. Koller M, Haenny P, Hess K, Weniger D, Zangger P (1990) Adjusted hypervolemic hemodilution in acute ischemic stroke. Stroke 21: 1429–1434
12. Kramer GC, Perreon PR, Lindsay DC (1986) Small-volume resuscitation with hypertonic saline dextran solution. Surgery 2: 239–247
13. Kuroiwa T, Shibutani M, Okeda R (1988) Non hyperemic blood flow restoration and brain edema in experimental focal cerebral ischemia. J Neurosurg 70: 73–80
14. Nakayama S, Kramer GC, Carlsson RC, Holcraft JW (1985) Infusion of very hypertonic saline to bled rats: membrane potentials and fluid shifts. J Surg Res 38: 180–186
15. Mast H, Marx P (1991) Neurological deterioration under isovolemic hemodilution with hydroxyethyl starch in acute cerebral ischemia. Stroke 22: 680–683
16. Messmer K, Kreimeier U (1989) Microcirulatory therapy in shock. Resuscitation 18: 51–60
17. Nolte D, Bayer M, Lehr H-A, Becker M, Krombach F, Kreimeier U, Messmer K (1992) Attenuation of postischemic microvascular disturbances in striated muscle by hyperosmolar saline dextran. Am J Physiol 263: H1411–H1416
18. Prough DS, Whitley JM, Taylor CL, DEal DD, DeWitt DS (1991) Regional cerebral bloodflow following resuscitation from hemorrhagic shock with hypertonic saline. Anesthesiology 75: 319–327
19. Schell RM, Cole DJ, Schultz RL, Osborne TN (1992) Temporary cerebral ischemia. Effects of pentastarch or albumin on reperfusion injury. Anesthesiology 77: 86–92
20. Seida M, Tomida S, Wagner H, Ting P, Klatzo I (1987) In: Cervos-Navarro J, Ferszt R, Role of reactive hyperemia in the dynamics of postischemic brain injury (eds). Stroke and microcirculation. Raven, New York, pp 277–280
21. Schürer L, Dautermann C, Härtl R, Murr R, Berger S, Röhrich F, Messmer K, Baethmann A (1992) Treatment of hemorrhagic hypotension with hypertonic/hyperoncotic solutions: effects on regional cerebral blood flow and brain surface oxygen tension. Eur Surg Res 24: 1–12
22. Strecker U, Dick W, Madjidi A, Ant M (1993) The effect of the type of colloid on the efficacy of hypertonic saline colloid mixtures in hemorrhagic shock: dextran versus hydroxyethyl starch. Resuscitaion 25: 41–57
23. Velasco IT, Pontieri V, Rocha e Silva M, Lopes OU (1980) Hyperosmotic NaCl and severe hemorrhagic shock. Am J Physiol 239: 664–673

Correspondence: Oliver Kempski, M.D., Institut für Neurochirurgische Pathophysiologie, Klinikum der Johannes Gutenberg Universität Mainz, D-55101 Mainz, Federal Republic of Germany.

Acta Neurochir (1996) [Suppl] 66: 118–121

Conduct of Head Injury Trials in the United States: The American Brain Injury Consortium (ABIC)

A. Marmarou

Medical College of Virginia, Richmond, VA, U.S.A.

Summary

This article provides a brief description of the American Brain Injury Consortium (ABIC), a group of clinical scientists focused on the design and implementation of clinical trials in the United States to improve outcome of the brain injured patient. The structure of the consortium, guidelines for participation, relationship with pharmaceutical industry, and the instruments used for acquisition and analysis of data gathered are described.

Keywords: Clinical trials; head injury; American Brain Injury Consortium (ABIC).

Introduction

The potential to improve outcome of patients with severe head injury with pharmacologic intervention was spurred by the results of the recent spinal cord injury trials indicating improved outcome with early administration of high dose steroids [1]. This trial was jointly funded by the National Institutes of Health and the pharmaceutical industry and until recent years, it was common that investigators would seek primary or partial funding for clinical trials from the federal government. This was the case several years ago when most if not all head injury trials were funded exclusively by the NIH. For example, clinical trials conducted by the Medical College of Virginia in Richmond, funded in this manner, included studies of barbiturate coma[4], Tromethamine (THAM) [2], and most recently, pegulated superoxide dismutase (PEG-SOD) [3]. However, the shrinking pool of research funds from the federal government coupled with a shift in policy toward federal funding of pharmaceutical product testing moved clinical investigators to the private sector for conduct of clinical trials. A new strategy was required to deal with the many problems facing the clinical investigator in the industry-university relation-ship where primary differences in mission, policy differences in freedom of information, ownership of data, publication rights and reporting of scientific findings play a role.

The Forming of the American Brain Injury Consortium

In April of 1993, a small group of clinical scientists met in Atlanta, Georgia, where a plan for organizing the head injury investigators in the United States was presented. The plan called for the formation of a consortium with the primary goal of designing and implementing clinical trials in head injury to improve patient outcome. Administratively, it was kept as simple and practical as possible with participation voluntary by all hospitals and university centers. Participation required only a pledge that the center investigator follows to the best of his ability the scientific guidelines set forth by the consortium. At present, the consortium consists of 60 Primary trauma centers throughout the United States.

The Administrative Structure

The Consortium administrative structure consists of a chairman, technical director, principal investigator, and center investigators (Fig. 1). All decisions by the Consortium are made democratically by the center investigators and each center investigator is entitled to one vote. The *chairman* is selected by nomination and approval by the center investigators attending scheduled Consortium meetings and serves for a minimum term of 2 years.

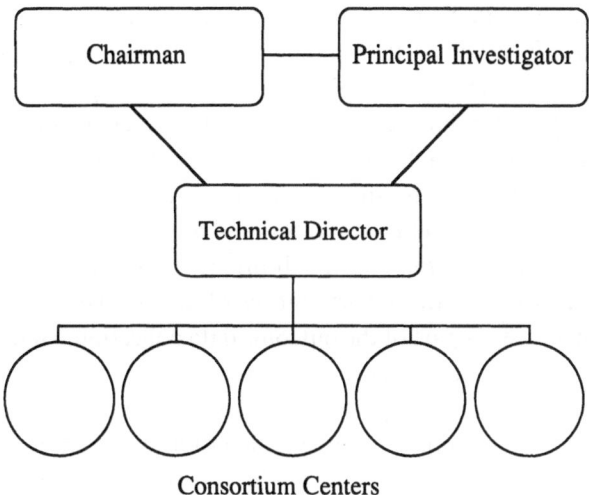

Fig. 1. Administrative structure of the American Brain Injury Consortium (ABIC). At present, there are 60 trauma centers in the United States and 7 Israel centers participating in ABIC

The *chairman's* responsibilities are: to represent the Consortium in all matters and as the official to be contacted by groups interested in utilizing the expertise and services of the Consortium, chairs all gatherings by the Consortium and maintains responsibility for organizing and implementing the meeting agenda, and assists the principal investigator and technical director in coordinating efforts with outside groups.

The *technical director's* responsibilities are: to assist the chairman, principal investigator and Consortium members in all technical aspects of the proposed study including: communication, data collection, processing, analysis and quality control; and to represent the consortium to outside agencies for all technical matters relating to clinical trial design and implementation.

The *principal investigator* of a clinical trial adopted by the Consortium must be a member of the Consortium. Usually, the center investigator has been engaged in experimental or clinical studies and has developed a close liaison with the pharmaceutical company for pre-clinical evaluation of their product. It is the responsibility of a potential principal investigator to introduce the proposed study to the consortium members and present evidence supporting the merit of the study design. If selected and approved by the Consortium, the principal investigator proposes scientific direction for the conduct of the study; coordinates scientific interchange between the Consortium and Agency; and accepts responsibility, as the first author, for reporting results of the study to the scientific community.

The responsibilities of the *center investigators* are: to decide which studies are to be implemented based on

the merits of the scientific data supporting the proposed clinical trial; participate in all scientific and policy decisions of the Consortium; conduct clinical trials at their respective center according to the guidelines set forth by the Consortium; propose or participate in sub-studies which may be warranted in conjunction with the main clinical trial; and participate in special committees established by the Consortium for management of centers, developing protocol clinical guidelines consistent with standards of care, contractual matters and publication.

Publication Guidelines

This is perhaps the most important problem facing the Consortium, particularly when it may not be in the best interest of the outside funding agency to report negative findings. However, the guidelines of the ABIC must be firm in this regard in order to be accepted not only by the center investigators but also the institutions they represent. The publication guidelines of the ABIC are straightforward. Simply stated, the Consortium maintains its right to publish and disseminate information resulting from the clinical trial following guidelines set forth by the Consortium, in cooperation with the funding agency, and in accordance with the rights of the individual participating centers. As stated earlier, the principal investigator is charged with the responsibility of writing the final manuscript and is designated senior author followed by the two center investigators contributing the most patients to the trial and "the ABIC". The ABIC members will be listed in alphabetical order in the appendix of the article. It is also important for the Chairman to designate a publications committee, which together with the principal investigator will oversee all information submitted for presentation or publication. In the process, the Consortium includes close communication with the funding agency to allow review and recommendations prior to publication.

Communication between Consortium Centers

The technical center of the ABIC resides in the Medical College of Virginia, which houses several computers for use by the Consortium. To facilitate a close communication with all centers, a computerized network was developed for instant transfer and receipt of information. The communication network provides a link either through conventional phone modem direct to the ABIC computer or by the Internet. Users with access to a university computer can use the "internet" as this only requires the cost of a local phone call. Those distant from a computer can purchase a commercial dial-in-service for connection to the "Internet". At present, both systems, modem connection to ABIC or Internet, are used for communication. However, as time progresses, all centers will be "Internet" based thus facilitating information transfer not only to the U.S. but worldwide.

Data Instruments:
The ABIC Clinical Research Forms

The main data collection instrument for most clinical trials is the hand written documentation on clinical research forms (CRF's). One of the major frustrations is that these forms vary considerably, and each pharmaceutical company has their own set. One task of the ABIC technical center was to develop a uniform set of CRF's which could be used for all clinical head injury trials and which would become the standard for the Nurse Coordinators at the respective centers. This was accomplished. After careful review of the initial forms introduced by the Traumatic Coma Data Bank and a specific "Neuroscore" set currently used by the MCV, a proposed draft of CRF's was submitted to the Consortium for review and, after refinement, adopted for use. These include forms for consent, initial screen, vital signs, ECG, injury information, GCS, pupillary response, CT-scan, multiple injury scales, medical history, medication history, laboratory data, study drug administration, outcome, study end, surgery treatment, narrative, concurrent medications, adverse events and serious adverse events.

These forms are computerized and reside on an "Ingress" relational data base for transfer to the funding agency, usually in SAS format, when quality control procedures have been completed and the final data set is "locked". Use of these forms will greatly simplify the implementation of clinical trials as they limit the amount of information collected while at the same time providing the data vital to the finding agency and the ABIC. Computerized forms for direct transmission to the ABIC are now in development for future application.

Scientific Sub-Projects Coordinated
within the Clinical Trial

An additional advantage of the ABIC is that scientific questions can be addressed, which can be incorporated into the clinical trial, which although not always useful to the funding agency is important for scientific purposes. This requires careful negotiation with the funding agency as every data element incorporated into the CRF's is subject to quality control and checking of source documentation, thus adding to the burden of the clinical monitors and to those processing data. One example of a sub-project that is presently underway is the incorporation of neuropsychologic measures as part of the outcome data collection. Heretofore, the use of neuropsychologic measures has been limited in this population because of the problems inherent in applying conventional tools to the severely injured patient. Thus, the ABIC formulated a short series of tests termed the "mini battery" as a complement to standard Glasgow outcome scores to provide additional measures of brain function. Each neuropsychological test selected serves as a useful indicator of individual cognitive functions including memory, attention, visual perception, motor functioning and language. The fundamental hypothesis is that treated patients will show better performance on measures of visual perception, visual memory, oral fluency, sustained attention, fine motor dexterity and quality of life.

Those tests which comprise the "mini-battery" along with information as to areas of measurement include; Rey Complex Figure Test (visuoconstruction and memory); Controlled Oral Word Association Test (oral fluency); Symbol Modalities Test, oral version (sustained attention); Grooved Pegboard Test (fine motor dexterity) and Neurobehavioral Functioning Inventory (Quality of Life). The time required to administer the tests ranges from 20 minutes for Good/Moderate outcome patients to 45 minutes for the severely disabled patients. This battery has been adopted by the ABIC and is in current use. It will provide the ABIC the opportunity to test the applicability of these neuropsychological and quality of life measures to clinical trials involving the severely injured patient.

Conclusion

This article briefly summarizes the construct and principles of the ABIC which was formed to provide a scientific body to conduct clinical trials in the private sector while at the same time maintaining academic standards of excellence. As care providers assume the ultimate responsibility for patient outcome, who is better qualified to decide how clinical trials involving new therapies or agents should be designed and implemented? At present, the ABIC consists of 60 premier trauma centers within the United States and an additional 7 centers in Israel. Shortly after the formation of the ABIC, all data collection instruments, methods of operation, and specialized programs were transferred to Glasgow. This eventually led to the formation of the European Brain Injury Consortium (EBIC). The EBIC has adopted these principles and together with ABIC and our pharmaceutical colleagues, will share the same problems and hopefully impact favorably on the outcome of the severely brain injured patient.

References

1. Bracken MB, Shepard MJ, Collins WF, Holford TR, Young W, Baskin DS, Eisenberg HM, Flamm E, Leo-Summers L, Maroon J, Marshall LF, Perot PL, Piepmeier J, Sonntag VKH, Wagner FC, Wilberger JE, Winn HR (1990) A randomized, controlled trial of methylprednisolone or naloxone in the treatment of acute spinal-cord injury. N Engl J Med 332 20: 1405–1411
2. Muizelaar JP, Marmarou A, Ward JD, Kontos HA, Choi SC, Becker DP, Gruemer H, Young HF (1991) Adverse effects of prolonged hyperventilation in patients with severe head injury: a randomized clinical trial . J Neurosurg 75: 731–739
3. Muizelaar JP, Marmarou A, Young HF, Choi SC, Wolf A, Schneider RL, Kontos HA (1993) Improving the outcome of severe head injury with the oxygen radical scavenger polyethylene glycol-conjugated superoxide dismutase: a phase II trial. J Neurosurg 78: 375–382, 1993
4. Ward JD, Becker DP, Miller JD, Choi SC, Marmarou A, Wood C, Newlon PG, Keenan R (1985) Failure of prophylactic barbiturate coma in the treatment of severe head injury. J Neurosurg 62: 383–388

Correspondence: Anthony Marmarou, Ph.D., Medical College of Virginia, P.O. Box 980508, Richmond, VA 23298–0508, U.S.A.

Index of Keywords

Acidosis 8, 56
Adhesion receptors 27
Albumin 15
American Brain Injury Consortium
    (ABIC) 118
Anatomical distribution 21
Antioxidant 107
Apoptosis 21
β-APP expression 96
Arachidonic acid 56
Axolemma 81
Axonal damage 81

Behavior 68
Brain plasticity 63

$^{45}Ca^{2+}$ 15
Cerebral infarction 40
Cerebral ischemia 1, 50, 68, 114
Clinical trials 118
Cytokines 40
Cytoskeleton 81
Cytotoxic brain edema 56

Deposition of Aβ 96
Disparity analysis 44

Environment 63

Excitotoxicity 50
Experimental brain infarcts 63

Facial nucleus 103
Focal cerebral ischemia 21
Free radicals 8
Freeze-lesion 15

Gene expression 1
Glia 103
Glial cells 56
Glucose metabolism 44
Glutamate release 50
Growth factors 103

Head injury 87, 96, 114, 118
Hemorrhagic infarction 76
Hippocampus 1

Image-processing 44
Immediate early genes 103
Immune reaction 32
Inflammation 27
Inflammation of the brain 40
Inflammatory changes 32
Intracranial pressure 114
Ischemia 8, 27, 107
Ischemic microvascular derangement 32

Leukocytes 27
Local cerebral blood flow 44

Macrophages 87
Microdialysis 50
Microglia 32, 87
Mitochondria 8
MK-801 73
Motor function 63

NBQX 73
Neocortex 68

Programmed cell death 1
Pyrrolopyrimidine 107

Reperfusion 76

Secondary brain damage 15
Small volume resuscitation 114
Stress protein 40
Stroke 27, 76

Temporal profile 21
Thrombolysis 76
Tirilazad 107
Transplantation 68
Traumatic brain injury 81

4-vessel occlusion 73

SpringerNeurosurgery

U. Ito, A. Baethmann, K.-A. Hossmann,
T. Kuroiwa, A. Marmarou, H.-J. Reulen,
K. Takakura (eds.)

Brain Edema IX

Proceedings of the Ninth International Symposium, Tokyo, May 16-19, 1993

1994. 281 partly coloured figures. XV, 590 pages.
Cloth DM 330,–, öS 2310,–
Reduced price for subscribers to "Acta Neurochirurgica":
Cloth DM 297,–, öS 2079,–
ISBN 3-211-82532-0
Acta Neurochirurgica, Supplement 60

This volume is an up-to-date report on progress in the understanding of brain edema, with a spectrum reaching from most recent molecularbiological findings to respective clinical developments. Major topics deal with (a) the blood-parenchymal cell border under normal and pathological conditions causing brain edema, (b) neuron-glial interactions and their disturbances in tissue damage, (c) formation, propagation and resolution of brain edema, and finally (d) treatment of vasogenic and cytotoxic brain edema. In the basic science approaches emphasis is given to newly discovered molecules, such as vascular endothelial growth factor, which might control permeability of the blood-brain barrier, e.g. in brain tumors. The complex issue of mediator compounds of secondary brain damage is further developed as to its manyfold involvement, for example in barrier dysfunction, cell swelling, disturbances of the microcirculation, and others. The report further contains comprehensive assessments of edema pathophysiology by advanced technologies, such as in-situ hybridization on the one hand side or NMR-diffusion imaging on the other. Novel forms of treatment acquiring increasing specificity represent a central focus.

SpringerWienNewYork

P.O.Box 89, A-1201 Wien • New York, NY 10010, 175 Fifth Avenue
Heidelberger Platz 3, D-14197 Berlin • Tokyo 113, 3-13, Hongo 3-chome, Bunkyo-ku